Techniques for Wildlife Investigations

Techniques for Wildlife Investigations
Design and Analysis of Capture Data

John R. Skalski
Center for Quantitative Science
in Forestry, Fisheries, and Wildlife
University of Washington
Seattle, Washington

Douglas S. Robson
Biometrics Unit
Cornell University
Ithaca, New York

Academic Press, Inc.
Harcourt Brace Jovanovich, Publishers
San Diego   New York   Boston   London   Sydney   Tokyo   Toronto

Academic Press, Inc.
1250 Sixth Avenue
San Diego, California 92101

United Kingdom Edition published by
Academic Press Limited
24–28 Oval Road, London NW1 7DX

Library of Congress Cataloging-in-Publication Data

Skalski, John R.
    Techniques for wild life investigations : design and analysis of capture data / John R. Skalski, Douglas S. Robson.
        p.   cm.
    Includes index.
    ISBN 0-12-647675-6
    1. Animal populations--Statistical methods. 2. Wildlife management--Research--Methodology. 3. Experimental design.
    I. Robson, Douglas S. II. Title.
    QL752.S52 1992
    333.95'411'072—dc20.   91-35261
                                                                    CIP

PRINTED IN THE UNITED STATES OF AMERICA
92 93 94 95 96 97   QW   9 8 7 6 5 4 3 2 1

Contents

List of Figures

List of Tables

Preface

This book brings together the topics of experimental inference and the use of mark–recapture data in wildlife investigations. Each of these topics is an important subject in its own right. We believe that despite the many unresolved issues in mark–recapture methodology, attention must be brought to focus on the use of this type of data in a hypothesis-testing framework.

In today's research environment, an interest in fish and wildlife population levels rarely ends with the estimation of animal abundance. Inevitably, there is a desire to compare estimates of abundance from different parts of a species' range or at different points in time. Similarly, decisions in resource management often require more than an inventory of wildlife populations. Important social and economic issues concerning the environment require timely answers to questions concerning the effects of human activities on wild populations. Yet, two decades after the National Environmental Protection Act (1970), the use of deductive inference in wildlife research is rare, and the primary emphasis on mark–recapture methods remains that of parameter estimation.

The emphasis of this book is on the design of field studies and the statistical inferences that can be made from observed changes in animal abundance. We focus our attention on how statistical inferences are influenced by the use of control populations, the spatial and temporal dimensions of a study, the selection and deployment of treatments, and the use of replicate populations in study designs. To these considerations we then add the ramifications of using mark–recapture methods in quantifying the population levels. The problem we address is how to design and analyze wildlife experiments when the response variable being measured, animal abundance, may be estimated with a coefficient of variance of 100% or greater.

When estimating animal abundance, we recommend model selection techniques such as those of Otis *et al.* (1978) for closed populations, and Cormack (1985) for open and closed populations as an essential first step in analyzing a wildlife investigation. We must leave it to our readers to incorporate the most appropriate estimator for their surveys into the tests of hypotheses we present. This same vast array of multiple-recapture models can also be used in the *post hoc* power calculations of wildlife investigations we present. However, prior to the field experiment, the anticipated trap response of animals is unpredictable, and the use of a complex survey model in sample size calculations is unfounded. Consequently, we have chosen to illustrate the design and analysis of wildlife investigations based on the parsimonious models of the Lincoln Index, the constant-probability Schnabel (1938) census [i.e., Model M_0 (Otis *et al.*, 1978)], and the constant-effort removal technique (Zippin, 1956, 1958). With these simple models, open population studies based on the Manly–Parr (Manly and Parr, 1968) technique and closed population studies subject to trap response [Model M_B (Otis *et al.*, 1978)] are explicitly incorporated. Also included are generalizations of the sample size calculations that can incorporate any survey model that possesses an explicit variance expression.

Fortunately, in comparative studies, the need for valid abundance estimation is not as important as is the need for valid indices of population change. We will show that estimation of animal abundance is not always necessary in analyzing mark–recapture data in a hypothesis-testing framework. Violations in model assumptions that would be serious in population estimation may be inconsequential when the object is detection of population change. Furthermore, we believe the data from a well-designed study will always lend itself to reanalysis using alternative survey models. In contrast, an experiment designed without proper regard to necessary replication, sampling effort, or experimental control will benefit little from whatever survey models are used in subsequent analyses.

The purpose of this book is to address two areas of methodology that are necessary to improve the ability of researchers to assess effects on wild populations: (1) the need for quantitative criteria useful in designing effective field experiments with reasonable probabilities of success, and (2) the need for valid statistical methods of analyzing mark–recapture data in a hypothesis-testing framework. Our intended audience includes biologists performing wildlife investigations, biometricians engaged in the design and analysis of field studies, and administrators required to evaluate the merits of ecological research. Readers of this book should appreciate that our stated goals will not be fulfilled by this published work alone. Both time and effort will be required in wildlife and statistical communities to bring the deductive framework of experimental ecology to the level held by such fields as industrial quality control, agriculture, and medicine. We limited ourselves to pursuing these goals within the context of the census techniques chosen. It is our hope that investigators will see the extension of these methods to other survey procedures such as density estimation, line-transect, and circular plot techniques.

In writing this book, issues have emerged that, although not vital to population estimation, are important in the design of multipopulation field experiments. Chapter 2 discusses the role of preliminary surveys in the design of field experiments and includes a discussion of three topics not often found in the wildlife literature. These topics are the estimation of variance components associated with spatial and temporal fluctuations in animal numbers, the use of cost functions in design optimization, and the modeling of catch-effort relationships for the prediction of trap performance. These issues likely will be topics of active research for years to come. Subsequent chapters incorporate these concepts in the design of abundance studies (Chapter 3), comparative censuses (Chapter 4), manipulative experiments (Chapter 5), and environmental assessments (Chapter 6).

Readers are expected to have a basic familiarity with mark–recapture methodology and the concepts of a statistical inference. It is assumed that readers have a basic understanding of parameter estimation, Type I and II experimental errors, confidence interval estimation, and the use of statistical tables. These requirements are usually satisfied by one or two courses in applied and mathematical statistics.

We would like to especially thank Mary Anne Simmons for the computer programming in support of this research. Our thanks is also extended to Kenneth Pollack and James Nichols for reviewing earlier versions of this book. Manuscript preparation was provided by Norma Phalen at Cornell University and Cindy Helfrich at the University of Washington; without their help, this work could not have been accomplished. Finally, we thank the U.S. Department of Energy (under Contract No. DE-AC06-76RL01830) for support of research that, years later, proved to be the inspiration for this book.

John R. Skalski
Douglas S. Robson

1

Statistical Inferences in Experimentation

The use of mark–recapture (animal marking and recapture) methods does not necessarily end with the estimation of animal abundance or density. Questions concerning natural or human-induced changes in animal abundance often require quantification. This quantification may begin with a series of mark–recapture studies but then needs to proceed to tests of hypotheses. Until recently, only two papers were published on methodology for comparing animal abundance in a hypothesis testing framework (Chapman 1951; Chapman and Overton 1966). This paucity of guidance has left researchers largely on their own and does not reflect the need for statistical tests of effects on mobile species.

The National Environmental Protection Act of 1970 established the need to assess the effects of construction activities and facility operations on local environments. Concerns about ecotoxicity and efforts to enhance wildlife abundance (through improved habitat management) also necessitate verification by experimental field tests. However, recent publications by Romesburg (1981) and Hurlbert (1984) admonish wildlife biologists for infrequently testing research hypotheses. They warn of the possible consequence of drawing inferences from field studies with little or no replication. Part of the problem may be that a clear distinction has not been made between the roles of sample surveys and field tests in wildlife investigations. Perhaps this is because the objectives of wildlife research often become confused with the limitations of population studies.

Some populations do not readily lend themselves to experimentation. Populations of endangered species and animals with large home ranges may not be appropriate for experimentation. In such instances, sample surveys and demographic studies may be the only approach to assessing population effects. However, other populations, including many prey species, game animals, and avian populations, can be studied and effects on their abundance detected in a hypothesis testing framework. The techniques in this book are developed for these latter species. In discussing experimental techniques, the distinctions and limitations of manipulative and survey studies will be addressed to help clarify their respective roles in wildlife research.

Some of the same reasons that have compelled researchers to use mark–recapture techniques in single-plot, population surveys also exist in field experiments and impact studies on mobile species. A complete census of wild populations is often impossible or cost-prohibitive; therefore, mark–recapture methods provide a useful technique for estimating the abundance of species with elusive or secretive behavior. However, the objectives of abundance estimation and hypothesis testing are different, as are the design requirements for these studies. A study that fails to take into account these design differences is likely to be inefficient and/or ineffective. For instance, the validity of mark–recapture models is essential in abundance estimation but is not necessarily so in comparative studies. In contrast, the pattern of animal dispersion across the landscape and the homogeneity of capture rates between populations are critical issues in multiplot, mark–recapture experiments but unimportant in population estimation. To efficiently design field experiments, researchers must take into account methodology specifically derived for the purposes of hypothesis testing.

Currently, index methods are used most frequently in field studies attempting population comparisons. Skalski et al. (1983a, 1984) showed that a catch index indeed may have a smaller sampling error than an estimate of absolute abundance, and has the potential of improving the precision of population comparisons. The use of such index data typically assumes a constant proportionality exists between the index that is measured and the animal abundance of the populations studied. Unfortunately, many wildlife surveys that yield index data do not have the fine structure necessary to test this assumption of constant calibration or homogeneity. Consequently, a primary purpose of mark–recapture methodology in tests of effects is to verify the assumption of homogeneity and, if the assumption is incorrect, to calibrate the catch data to absolute abundance estimates in order to provide valid treatment comparisons. Statistical tests presented in this book support this suggestion and point to the need for mark–recapture as a fail-safe feature in field studies designed for hypothesis testing. Thus, mark–recapture methods have a role in hypothesis testing independent of their traditional role in abundance estimation.

Scope of This Book

In this book, we discuss the various considerations necessary when incorporating mark–recapture methods into population investigations that have as their objective abundance estimation, comparative censuses, test of treatment hypotheses, or the assessment of environmental impacts. Within each of these categories of investigation, the design, analysis, and interpretation of mark–recapture data will be addressed. The necessity of envisioning the investigatory process as an unbroken flow from the statement of the research objectives to the design, analysis, and interpretation of results will be stressed (Figure 1.1). Nowhere is the integration of this process more important than in population investigations where the magnitude of error variances can be large, potential for confounding effects likely, and costs of investigation great.

The development of this mark–recapture methodology will begin with a discussion of the principles of experimentation that might be considered unnecessary if it were not for the fact that population studies often fail to incorporate crucial design concepts (Chapter 1). Ronald A. Fisher (1947, p. 34) has stated that when a study fails to provide a valid estimate of experimental error, it is more an experience than it is an experiment. The need for control populations and the use of randomization and replication of treatments to provide valid inferences to effects on population abundance will be discussed. The distinction between observational studies, manipulative experiments, and impact assessments will be presented in terms of their design principles and the statistical inferences sought.

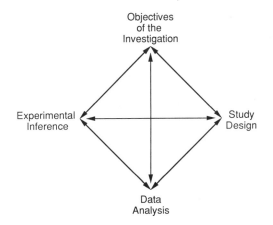

Figure 1.1 Interrelationship between research objectives, study design, data analysis, and statistical inference in field investigations.

The design and analysis of preliminary sample surveys of wild populations using mark–recapture methods are presented in Chapter 2. Chapter 2 shows how preliminary sampling can be used to provide the prior information necessary for the design of effective field studies. An irony, seemingly in the design of field studies, is that efficient and effective experiments can be designed only once the probable outcome of the study is known. The chapter includes methods for estimating the variance components associated with multiplot, mark–recapture studies and the spatial heterogeneity of wild populations. There are also discussions of catch–effort models (e.g., relationship between trapping effort and probability of capture), variance relationships, and cost functions. We suggest that readers become familiar with these concepts before proceeding to the later chapters, which incorporate these concepts in the design and optimization of field studies.

Chapters 3 through 6 are devoted to the design and analysis of population investigations using mark–recapture methods. The focus of Chapter 3 is on the design of abundance studies and the statistical analysis of capture data after animal abundance has been estimated. Whereas most of the literature on mark–recapture studies has addressed the subject of abundance estimation [for reviews, see Cormack (1968, 1979) and Seber (1982)], little guidance is available on the choice of study plot size, level of trapping effort, and survey duration. Using the results of Chapter 2, approaches to the optimal design of abundance studies will be presented. Aspects of optimizing the design of abundance studies will be used later in the design of multiplot field studies in Chapters 4–6.

In Chapter 4, comparative censuses will be examined. Here, statistical methods for the design and analysis of field studies with the objective of comparing animal abundance between two locations in a species range or through time are presented. The results of this chapter will be used in the subsequent design of paired experimental approaches in Chapters 5 and 6.

The design and analysis of manipulative experiments to test effects on animal abundance are presented in Chapter 5. These designs employ the properties of randomization, replication, and blocking to provide valid tests for differences in mean abundance between control and treated populations. We show that the form of the test statistic depends on the degree of homogeneity in capture probabilities among the populations compared. Power functions for these tests of treatment effects are used in the design of field experiments. Application of open- and closed-population models to field experiments is discussed.

In Chapter 6, the effects of restricted randomization and replication on the inferences of field investigations will be addressed. Such restrictions can occur in impact assessments where field biologists have limited control over the waste process flow or in investigations confined to limited geographic areas. Three designs will be presented that are useful in tests of impact hypotheses. One design has inferential value when randomization is restricted; the other two designs can be used in the absence of both spatial replication and randomization.

Objectives of Population Investigations

According to Das and Giri (1979, p. 1), experimentation and the formulation of inferences are the twin essentials of the scientific method. Nevertheless, observational studies form the majority of today's population investigations. The intended purposes of observational studies and manipulative experiments make neither approach to population investigation inherently superior. However, there may be a distinct disadvantage in applying the wrong investigative style to a particular objective. In wildlife investigations, the option of a manipulative experiment does not always exist; and an observational or environmental assessment study may be the only alternative. The goal is to identify the most efficient and effective study design for the objectives within the constraints of the research.

Observational Studies

The principle use of mark–recapture methodology has been for parameter estimation in descriptive or observational studies. Surprisingly, however, survey sampling methodology has not been adapted to include the use of mark–recapture methods for the purposes of estimating mean abundance or inventorying large contiguous populations. Instead, mark–recapture methods have typically served to estimate abundance at a single study plot. The statistical inference in this case is limited to the animal abundance at that study area.

In an observational study, a sample is taken from a finite collection of subjects. These subjects may be the individual animals of a population, or a series of wild populations. The general objective of an observational study is to estimate parameters useful in describing the statistical population. In wildlife investigations, this might include the estimation of animal abundance (or density), survival rates, natality rates, or age and sex structure of a population. Such investigations often are called "descriptive" studies when the objective is basically parameter estimation. When the parameter estimates are to be used to study the viability of a population or detect a change in population status, the investigation is often given the more specific name of a demographic study. The statistical inference in either case is limited to the collection of subjects from which the sample was drawn.

In many wildlife investigations, a statistical inference beyond the site boundaries is both desirable and necessary. In order to make this inference, the larger statistical population must be defined and sampled. A statistical population consists of all possible observations of the response variable of interest. A useful definition of a statistical population in wildlife research often will require both spatial and temporal characterization. Typically, the spatial dimension will be the focus of probability sampling, whereas the temporal dimension is systematically sampled during periods of demographic importance. For example, one might be interested in estimating the mean abundance of the least chipmunk (*Eutamias minimus*) in ponderosa pine (*Pinus ponderosa*) habitat of Oregon during fall 1991.

The first step in devising a sampling scheme is to construct a sampling frame. In surveying the chipmunk abundance, the sampling frame might consist of a listing of all possible five-hectare (5-ha) plots composed of ponderosa pine habitat in Oregon. At this point, the survey is confronted by the magnitude of the task. Sites within the listing would be numerous, widespread, and possibly inaccessible. A prudent reaction to these logistical problems might be to consider redefining the sampling frame. Possible considerations might include limiting the study sites to publicly owned lands or sites within a specific county or township of interest. Such a reaction would make the survey more feasible, but it also potentially conflicts with the stated objectives and target population.

A target population is the statistical population of inference. In the example, the target population is the fall chipmunk abundance in a ponderosa pine habitat in Oregon. For valid statistical inference to this target population, the sampling frame must match this target population. If the sampling frame is revised, a corresponding change in the target population is implicit. It is important that an investigator appreciate this corresponding change. Any inference to the initial target population will depend on the representative nature of this revised population and the judgment of the investigator, not on statistical findings. The principle is straightforward; a statistical inference can be made only to a population that has been representatively sampled according to the laws of chance.

Another important use of observational studies is in the characterization of processes or relationships among elements of a natural system. The subjects of the investigation are once again a sample from a single statistical population. Here, however, the self-assigned differences or traits in which the subjects may systematically vary are the focus of the study. To study a single causal influence, the investigator must try to adjust the observations for the effects of other confounding factors through regression analysis. Such adjustments typically make the statistical analysis more complicated and the conclusions less rigorous than in randomized studies. The reasons for the uncertainty lie in the fact that the measurements and adjustments are model-dependent and cannot be readily assumed to correct for all the important factors in which the populations may systematically differ. This is to be expected because the adjustments are based on simplified assumptions concerning the statistical relationships between explanatory variables and the measured response. As a consequence, the statistical inferences generally will be weaker than those of manipulative experiments that eliminate confounding through the physical process of randomization. The validity of inferences from regression analysis, as such, will depend not only on study design but also on the skills and ability of the investigator to develop sound subjective interpretations of the test results.

In sample surveys, properties of random selection and finite sampling theory are used to provide valid inferences to the statistical population. Based on the laws of probability, the chance that an atypical sample is selected can be computed, and the probability of making an erroneous conclusion is known. Sample survey methodology acknowledges the potential of an error in making generalizations from a sample, but, unlike personal judgment, the error rate can be objectively determined. It is this objectivity that is persuasive in statistical inference. Investigators may disagree about the appropriateness of a sampling frame and target population, but inferences to those populations generally will be accepted as sound if principles of sampling theory are followed.

Statistical inference from a mark–recapture study is of a special form, and consequently, its nature requires additional attention. Validity of the statistical inference will depend not only on adherence to the rules of probability sampling in surveys of wildlife resources but also on the specification of an appropriate survey model for the capture data. A survey model incorporates a mathematical representation for the trap response of animals so that animal abundance can be estimated from incomplete counts. In modeling the trap response, assumptions such as constant probability of capture, equal probabilities of capture for marked and unmarked animals, and homogeneous capture rates among individuals are incorporated into the abundance estimation techniques. Unfortunately, the validity of these assumptions and the appropriateness of a mark–recapture model cannot be determined *a priori* as in survey sampling methodology. Rather, the correctness of a model must be tested empirically after the data have been collected. As such, models for mark–recapture studies are at best good approximations to the truth, and statistical inferences to a specific population are only as good as the techniques used in model selection.

In the survey to estimate mean abundance of the least chipmunk, an investigator will be confronted with both variation in abundance across the landscape and sampling error associated with estimating animal abundance at the individual sites. When mark–recapture studies are used to estimate animal abundance at replicate study sites, the sampling error will be of a parametric nature whose form will depend on the model selected. In contrast, estimation of the spatial component of variation in animal abundance will be of a nonparametric nature, which is unaffected by the pattern of animal dispersion. This incorporation of a parametric model for sampling error into survey designs distinguishes studies of animal abundance from other investigations. As a consequence, an investigator must be concerned with both the design of the research and the validity of the underlying capture model when making statistical inferences to wild populations.

Manipulative Experiments

When the relative comparison of populations under varied conditions rather than absolute determination of a population characteristic under a given condition is sought, a manipulative experiment is often the most effective research approach. In field ecology, the term "manipulative experiment" has been used generally to include any field study that attempts to compare control and altered conditions of ecological systems. Other adjectives such as "planned," "controlled," or "comparative" have been used to describe such studies. Here the term "manipulative experiment" will be limited to field studies that use the principles of randomization and replication of treatment levels when testing effects on animal abundance. An important aspect of this experimentation is the creation or specification of control conditions. Treatment differences are then assessed relative to the magnitude of the variance among plots treated alike.

Cox (1958) gives three conditions that prompt an experimental approach to answering a research question: (1) an objective of comparing treatment effects, (2) a substantial variation in response from plot to plot in the absence of a treatment effect, and (3) treatment differences that are relatively stable despite possible fluctuations in mean response levels. Under these circumstances, a direct comparison of treatment conditions may be more efficient than a comparison of the mean response of separate statistical populations. In wildlife investigations, an experimental approach may be useful in tests of management practices or habitat change (LoBue and Darnell 1959; Tester 1965; Rosenzweig 1973; Gains et al. 1979), test of environmental toxicity (Barrett and Darnell 1967; Pomeroy and Barrett 1975), evaluation of population regulatory mechanisms (Chitty and Phipps 1966; Tamarin and Krebs 1972; Krebs et al. 1969), and testing for the existence of species competition (Joule and Jameson 1972; Schroder and Rosenzweig 1975; Cameron 1977; Munger and Brown 1981).

Inferences from manipulative experiments stem from the ability to create experimental conditions and to randomly select which wild populations will receive a given treatment. This investigative process assures that no systematic differences exist between populations except for differences between treatments. The result is an unambiguous comparison between treatments whose validity depends solely on how the study was conducted. Because experimental conditions can be repeatedly recreated, the inferences from a manipulative experiment are actually to the processes that generated the observations, and not solely to the experimental units used in the study. These processes are defined by site-selection criteria, field techniques used, and treatment conditions ascribed to the populations. In such experiments, the observations are simply a sample from an infinite number of possible realizations of the test conditions.

An experiment has been defined as the collecting of measurements or observations according to a prearranged plan for the purpose of obtaining evidence to test a theory or hypothesis (Federer 1973). Fisher (1947) would have added to this definition the necessity that the study provide a valid estimate of error variance with which to test hypotheses. This experimental error, or error variance, is a measure of the extraneous variation that tends to mask the effects of treatments. For instance, the correct measure of the error variance for testing treatment means is "one that contains all sources of variation inherent in the variation among treatment means except that portion of the variance due specifically to the treatments themselves" (Fisher 1947). In a manipulative experiment, randomization and replication of the treatment conditions are necessary prerequisites for providing a valid estimate of the error variance used in testing treatment hypotheses.

Construction of a hypothesis implies a belief that there exists a degree of order or regularity that can be identified and measured despite fluctuations in response. To statistically test a hypothesis concerning a population response, the hypothesis must be first translated into statistical terms. A statistical hypothesis is a statement or conjecture about the parameters of one or more statistical populations. In this book, tests of treatment effects will be concerned with the differences in mean abundance among replicate populations under various test conditions.

Choice of treatments in an experiment will have a major effect on interpretation and understanding of test results. A fundamental of experimentation is the ability to alter environmental conditions in order that treatment effects can be observed. The manner in which these factors are altered, however, will serve to define the scope of the research. The objective of a study may be simply to identify the better treatment or, in addition, to provide clues as to why treatments behave as they do. Single-factor experiments are useful in making a choice between alternative treatment conditions. When an understanding of interrelationships among causal factors is also important, two or more variables may be changed simultaneously to study their interactions. Factorial treatment designs (Kirk 1982, pp. 350–353, 429–441) provide a structure of treatment combinations which permits simultaneous tests of main effects and interactions between factors. Sometimes the understanding obtained from this basic research also leads to identification of preferred treatment conditions.

The range of statistical inference derived from treatment comparisons will depend on the sources of error included in treatment contrasts. In some experiments, the representative nature of the experimental units is not of great interest. In such situations, material selected for testing is specifically chosen for its uniformity and to present results in a simple illustrative form. The primary interest in such experiments is often simply in the insight obtained on mechanisms of treatment effects. The error term as a consequence will not generally include an adequate measure of normal heterogeneity between experimental units in order that an inference to a target population of wide interest be valid.

In population studies, initial tests may be conducted on very similar plots in a contiguous habitat to reduce variance in animal abundance and increase chances of detecting treatment differences. To understand the implications of the resulting tests, further work would be necessary under varied environmental conditions. In these latter studies, the objective would be to determine whether treatment effects remain stable from locality to locality or among habitats. Usually, a series of experiments is necessary to test any important hypothesis.

In research to support management decisions, the representative nature of experimental units is vital for valid inferences. The experimental conditions (i.e., application of treatments and experimental units) need to be as representative as possible of conditions under which the test results are to be applied. A wide range of conditions may be purposefully sought to ensure a broad basis of inference and to interpolate results to intermediate conditions. In such circumstances, different environmental conditions or habitats are selected to determine whether treatment differences remain relatively constant. It is this property of additivity that permits inferences to conditions not represented in statistical tests of treatment effects.

In still other circumstances, the objective of an experiment may be to make inferences to a specific target population. Here, the experimental units or study plots, as in the case of wildlife investigations, need to be a random sample of the larger population to which inferences are intended. As in observational studies, success of a manipulative experiment depends on properly identifying a target population and sampling it representatively.

Environmental Impact and Assessment Studies
In field research, the only options available may be to conduct a constrained study or not perform the investigation at all. Such conditions occur frequently in environmental assessment. Since the passage of the National En-

vironmental Protection Act of 1970, investigators have been compelled to perform tests of impact on natural environments and wild populations. Typically, in such investigations, there is only one nuclear power plant, one refinery operation, one building permit, or one accidental chemical or oil spill. Without replication and randomization of the treatment conditions, a manipulative experiment is not possible. Instead, a site-specific study may be attempted with the limited objective of population assessment in the immediate vicinity of the potential impact. Legal requirements and social interests often necessitate a site-by-site assessment even when replication may exist. Because of the importance of environmental assessment, investigators need to have a basic understanding of the unique characteristics and limitation of this type of field research.

Another field situation that can confront investigators is when treatment conditions are replicated but assignment of treatments to wild populations is outside the investigator's control. Examples include tests of possible population effects from high-voltage transmission lines, pesticide spraying programs, reforestation projects, and salting programs to control road ice. In each of these examples, the treatment condition is repeated across the landscape but is not randomly assigned to test plots. Rather, treatment sites will be chosen for their unique characteristics or their need for treatment. For instance, in tests of secondary poisoning in wildlife, plots of land that receive pesticide spray will be those, and only those, with infestation rather than a random selection of test sites. Thus, the property of randomization vital in testing treatment hypotheses is absent. Here again, a manipulative experiment cannot be conducted and the data cannot be used to test a treatment hypothesis. The only recourse is a test of impact taking advantage of treatment conditions that might otherwise go uninvestigated or be too costly to reproduce in an experimental setting.

The absence of randomization and/or replication distinguishes impact and assessment studies from manipulative experiments (Figure 1.2). An assessment study generally implies an investigation in the absence of true replication of the experimental unit, whereas an impact study generally implies a lack of randomization of test conditions. A field study may lack one or both of these design principles. The purpose of a test of an impact hypothesis is to determine the existence of a systematically or haphazardly applied effect on experiment units. This differs from a test of a treatment hypothesis where treatments are randomized to experimental units according to laws of chance.

In the absence of both randomization and replication of test conditions, only an impact assessment can be conducted (Figure 1.2). In an impact assessment of an electrical generating station, for example, the control conditions are repeated yearly during the preoperational phase of plant opera-

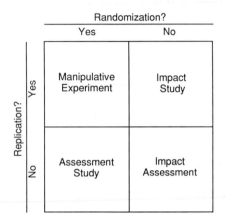

Figure 1.2 Relationship between design principles of randomization and replication and the nature of environmental field studies.

tions, then followed by a block of time under operational conditions. This systematic occurrence of test conditions in separate blocks of time precludes the randomization of effects. Furthermore, because there is only one experimental unit, the single power plant, inferences cannot be made to the population of power plants or their environmental effects, but only to effects at a specific site. Thus, the investigation is more correctly considered to be an impact assessment than a manipulative experiment. Failure to recognize the distinction may result in an inability to make sound inferences to even the site-specific effects.

Abandonment of essential design principles of randomization and/or replication in assessment and impact designs occurs with a price. Generally, the inferences from constrained studies will not be as rigorous as those from studies where test conditions have been experimentally applied to eliminate confounding with environmental effects. To enhance the validity of conclusions from impact and assessment studies, a temporal dimension often is added to the design of the studies. The purpose of sampling through time is often to eliminate extraneous effects with the aid of proposed response models. As such, not only are the design principles less understood, but environmental assessments must include design elements that can be avoided in manipulative experiments. In Chapter 6, we will show that these constrained field studies actually sacrifice some efficiency to enhance validity of their conclusions.

Design and analysis of an impact assessment will change depending on the extent of prior knowledge about the nature and source of the potential impact. As Green (1979) advocated, the best approach includes the use of both spatial and temporal controls to detect impacts on dynamic systems. To employ such a design, the location and timing of the potential impact must be known prior to its occurrence. In the case of accident assessment associated with oil and toxic waste spills, forest fires, and other natural disasters, "optimal" designs are seldom possible. Such tests of impact must depend on detecting spatial patterns correlated with the stressor or time-by-treatment interactions. Structure of the error variance will vary greatly depending on choice of the field design. In manipulative experiments, the error variance always includes a term for the spatial variance in animal abundance. In assessment studies, the error term can include spatial and/or temporal measurements of variation in animal abundance. Nowhere is the interrelationship between study objectives, field design, and statistical analysis more closely tied than in assessment studies where these factors can be so easily influenced by an ill-conceived investigation. Thus, investigators must be diligent in designing impact studies and devising the associated statistical tests for impact.

Design Considerations in Population Studies

Experimental Design Principles
A perfect study design requires a perfect image of nature (Churchman 1948) and would take into account all sources of variability associated with fluctuations in population numbers. In the absence of perfection, the best approach is to use a field design that takes into account all known sources of variation not directly associated with treatment differences. Randomization is then used to eliminate all other sources of systematic change. The complexity of a field experiment will depend to a large extent on the number of treatments and the number of sources of variation to be controlled. In manipulative experiments, there are generally only spatial sources of variation to control. In general, the greater complexity of impact studies is due to the need to control both spatial and temporal sources of variation that cannot be randomized between treatments. A reasonable rule is to use the simplest experimental design that provides adequate control of variability (Federer 1955).

Cox (1958) gave the following five criteria necessary for a good experiment: (1) elimination of systematic errors that may confound treatment effects, (2) precise estimates of treatment contrasts, (3) experimental conclusions with a useful range of validity, (4) simplicity in experimental design, and (5) ability to estimate an error variance. The best choice in an experimental design is one that provides the desired level of precision with the smallest expenditure of time and effort. A more complex design has little merit if it does not improve the performance of statistical tests or provide more precise parameter estimates. A usual side benefit from using an efficient experimental design is simpler data analysis and cleaner experimental inferences.

The three principles of *replication, randomization,* and *error control* given by Fisher (1947) are essential for estimation of the error variance and tests of treatment effects. Most field studies that fail do so because one or more of these basic principles were not incorporated into the experimental design. A discussion of these three principles as they apply to field experiments should be helpful in avoiding common pitfalls.

Replication   The level of replication needed in an experiment will depend on a number of interrelated design elements, including (1) size and shape of study plots, (2) animal density at sites, (3) heterogeneity of the landscape and its effect on animal dispersion, (4) dispersion pattern of study plots, (5) extent and nature of interplot competition, (6) cost of establishing study plots relative to other refinements of the experimental approach, (7) level of sampling at individual study plots, and (8) the particular experimental design chosen. These considerations need to be viewed concurrently. In multiplot mark–recapture studies, an experiment might have the optimal size and number of replicate study areas but be rendered ineffective because of insufficient trapping effort. Accurate determination of sample size generally requires some knowledge about the state of nature and anticipated response of the population to the treatment effects. Preliminary sampling may be used to acquire much of this information for the design of experiments (see Chapter 2).

The single most important parameter influencing the level of replication is the magnitude of the experimental error variance (σ^2_{EX}). A preliminary estimate of this error variance can often be derived from past experiments of similar design, higher-order interactions of factorial experiments, and theoretical considerations of the nature of the experimental unit, as well as from preliminary surveys. The more accurate this estimate and the better understood are the individual sources of error contributing to its overall variance, the more reliable the resulting sample size calculations. In later chapters, variance components will be used to optimize field study designs taking into account both the level of replication and sampling effort.

The well-known formula for the variance of a sample mean

$$\mathrm{Var}(\bar{x}) = \frac{\sigma^2}{n}$$

where n is the number of replicates, applies equally well to wildlife studies where animal abundance must be estimated. By increasing the level of replication, the magnitude of the variance of treatment contrasts is decreased and the precision of the field study improved.

When each treatment receives the same level of replication, the experimental design is said to be balanced. Balanced designs tend to not only simplify the data analyses but also simplify the interpretations of test results. In complex experimental designs with nested and cross-classified experimental units, a balanced design is almost essential for clear interpretation of statistical tests [see Searle (1971, pp. 316–318), on the interpretation of an unbalanced two-way cross-classified design].

In studies to test effects on wildlife abundance, balanced designs are important for yet another reason. Within-treatment variances will tend to vary when the mean abundance of wild populations differs as a result of treatment effects. The probability distributions of many test statistics, while assuming homoscedasticity, remain reasonably well-behaved when the treatments have unequal variances and the designs are balanced. For the purposes of data interpretation, balanced designs are strongly preferred in wildlife abundance experiments where treatment variances can change greatly (Skalski 1985a).

A potential side effect of increasing the level of replication is a change in the spatial proximity of the study sites. Adding sites to the field experiment will generally require that the experiment be conducted over a greater area. The variance among plots treated alike, in turn, can be expected to increase during a design process intended to decrease the overall variance. As such, the variance of mean abundance (i.e., the precision of the study) may not be a strict linear function of $1/n$. Chapter 2 provides guidance on balancing the competing effects of increasing sample size and greater interplot variation on the overall variance of the experiment.

Another important consideration in plot dispersal is the requirement of independent observations. In tests involving manufactured goods, the experimental units are often physically distinct and so the assumption of independence is readily satisfied. In field studies, however, the study plots are areas within a continuous landscape making physical separation difficult and the assumption of independence less straightforward. Two forms of competition that may violate the assumption of independence are intraplot competition and interplot competition. Intraplot competition can occur when the trapping devices at a plot do not act independently. This can happen when there are too many trapping devices and the animals avoid capture because of the site disturbance they cause. Too few trapping devices can result in the opposite effect where the animals "compete" with each other to be captured. Competition within an experimental unit is typically not a problem if it represents conditions under investigation, or if it occurs within each unit and the variance among plots is used to test treatment hypotheses.

In contrast, the presence of interplot competition is always of concern. "Interplot competition" refers to deleterious or advantageous effects of one experimental unit on another (e.g., movement of animals from one plot to another). Deleterious treatment effects may cause animal movement from treated areas to control sites, or depleted sites may recruit immigrants from plots with higher abundance. Borders, guard material, or use of control strips of sufficient distance (minimum of one home range) may be necessary to reduce such movements. However, the use of fenced enclosures has been shown to influence the demographics of wild populations (Krebs et al. 1969, Boonstra and Krebs 1977). In general, it appears that spatial separation is preferable to the use of enclosures in the design of wildlife field experiments.

The sequence in which study sites are processed also can introduce interplot competition by producing a positive correlation between sites. This often occurs when it is easier or cheaper to process all replicates of one treatment before proceeding to the next treatment. For instance, in field studies, systematic trap mortality may occur if all plots of a particular treatment are checked later in the day than those of another test condition. For easier data analysis, the same restrictions should be used in randomizing the handling scheme as were used in assigning treatments to plots. If a completely randomized experimental design is employed, the plots should be processed in a random order. A paired or randomized block design would suggest using a handling scheme based on that same blocking design.

Randomization   Any system of allocating treatments to study sites that renders the treatments more alike or unlike than expected by random allocation will result in an estimate of the error variance that is too large or too small, respectively. Similarly, this allocation scheme will also produce treatment differences that are smaller or greater than expected. To provide unbiased treatment comparisons and estimates of error variance, treatments need to be assigned to plots according to the laws of chance within the constraints of the experimental design.

Randomization does not guarantee even distribution of the natural differences among study sites between treatments. Rather, it is a way to ensure that a treatment is not consistently favored or impaired by extraneous or unexpected sources of variation. Even if an unexpected source of variation is eventually recognized, its effect has been randomized, and in so doing, still provides unbiased estimates of treatment contrasts and the error variance. Hence, the process of randomization is necessary to guard against unknown sources of experimental bias. Randomization eliminates the problem of trying to guess which biases might occur, and defending the position that all important sources of error have been controlled.

When randomization is properly used, each specific randomization of treatments has an equal chance of occurring. Consequently, each treatment has an equal chance of being favored by a particular randomization. It is this chance process that permits the significance level of a test to be computed. Furthermore, as the number of replicates used in an experiment increases, the chance of an unfavorable randomization decreases. Wilson (1952, pp. 46–48) presents a contrived example that shows how increased levels of replication decrease the likelihood of mismatching treatments to plots that have striking differences in response level.

In animal population studies, the level of replication is likely to be small to moderate in number. In such situations, the chance of hitting on a randomization scheme that seems particularly unsuitable is reasonably likely. A choice then has to be made whether to use this extreme randomization pattern. Laws of probability indicate that in the long run, extraneous factors will be averaged out of the treatment comparison. However, for a particular experiment, the objective is to conduct a well-designed study and to draw useful conclusions. Cox (1958, pp. 85–86) presents an excellent discussion of the philosophical and practical consequences of this dilemma. He suggests three approaches to resolving the problem of extreme randomization patterns in small experiments: (1) restrict the randomization process so that extreme patterns cannot occur, (2) establish a rejection rule prior to the randomization that applies equally to all treatments, and (3) select the randomization from a special set of suitable randomizations. A restricted randomization of experimental units will usually be preferable to using a field design that knowingly results in unfavorable treatment comparisons.

Error Control and Reduction   Several standard means by which the precision of an experiment can be improved include (Cochran and Cox 1957; Cox 1958): (1) greater use of experimental controls, (2) a larger experiment with greater replication, (3) refinement of the experimental techniques including greater sampling precision within experimental units, and (4) improved experimental designs including the use of blocking and covariates. To varying degrees, all of these practices have a role in the design of wildlife experiments.

The use of control measures must always be tempered by the understanding that the conditions of an experiment are likely to become less realistic as control measures increase. "Control measures" are defined as those experimental procedures that are used to minimize or eliminate sources of variability (e.g., environmental growth chambers are used to precisely control climatic conditions). Laboratory colonies of animals often provide excellent control of extraneous sources of variation that may influence test results. However, inferences from such studies to natural populations often are tenuous and require additional field verification. Experience often will dictate which factors need to be controlled, which can be ignored, and which other factors must be incorporated into field designs.

By using pairing or blocking in field designs, investigators attempt to assign as much of the heterogeneity in experimental material as possible to differences between blocks. In so doing, one can eliminate that portion of the variation from the experimental error. The object of blocking (or pairing) is to make individual plots as similar as possible within a block, while at the same time making the blocks as dissimilar as possible. Populations in a block should give as nearly as possible identical abundance estimates in the absence of treatment effects. Although blocking will always eliminate systematic errors, the extent of the error reduction will depend on the skill of the investigator in identifying similar sites. A reasonable starting point is to select sites that are in close proximity and conduct wildlife surveys close together in time. Vegetation surveys and on-site reconnaissance can provide additional data useful in improving the pairing of study sites (Chapter 2).

Effectiveness of blocking generally will decline as the number of treatments to be tested increases. The reason is that the physical size of the block will increase, causing the study plots to become more dispersed and dissimilar. Blocking is particularly useful in wildlife field experiments where animal abundance can change greatly across habitats and localities. By such use of blocking, experimental inferences to a range of environmental conditions are possible without drastically increasing experimental error.

Federer (1955, p. 69) gives a useful illustration of the construction of blocks and choice of plot shape in the presence of environmental gradients. From his example, it is clear that blocks should be as compact and small as possible (preferably square-shaped) and positioned along an existing environmental gradient. In this way, each block will be as dissimilar as possible and the area within the block will be as homogeneous as possible. Within blocks, however, Federer (1955) shows that plots should be rectangular in shape with their length running parallel to the direction of the gradient (Figure 1.3). In this configuration, plots within blocks will be as similar as possible and each plot will possess the same changes along the gradient.

Skalski (1985a; Skalski et al. 1984) showed that field designs that promote homogeneous capture probabilities between populations will result in comparisons with smaller sampling error. Blocking on time is one way to enhance prospects of homogeneity when all populations cannot be surveyed simultaneously. The same blocks used in grouping the study plots also should be used in the scheduling mark–recapture surveys to take total advantage of the error-reduction properties of blocking.

In population studies where abundance is going to be estimated using mark–recapture techniques, refinement of experimental procedures can substantially contribute to error reduction. Unbiased estimates of abundance can improve both accuracy and interpretation of treatment comparisons.

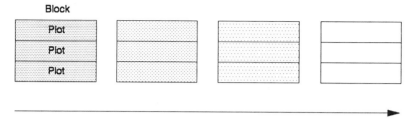

Figure 1.3 Preferred orientation and shape of blocks and study plots within blocks when in the presence of an environmental gradient. Arrangement maximizes the difference between blocks and minimizes differences between plots within blocks.

Treatment Design Principles

Goals of an investigation are achieved largely through the selection of treatments incorporated into the experiment. A "treatment" is defined simply as a procedure whose effects are to be measured or compared during the course of the experiment. Selection and arrangement of these treatments in an experiment is referred to collectively as the "treatment design." An important consideration in the selection of treatments is the possible effect of reduced replication in allowance for a greater number of treatment levels.

An important aspect of a manipulative experiment is the specification of control conditions to be tested. An experimental control group can be regarded as a treatment in which one is not particularly interested but that is necessary in quantifying the effects of other treatments. Controls may represent baseline conditions, a standard treatment for comparison with a new process, or a carefully regulated condition that is similar to the other treatment processes except for the variable of interest. To say that a mean abundance was observed previously and that a new treatment now gives a different value is not sufficient evidence of a change. Control and treatment conditions must be compared concurrently with sufficient replication to make a statistical inference concerning a difference in mean animal abundance.

A final prerequisite for a successful experiment is the presence of conditions necessary for expression of treatment effects. In wildlife investigations, an elaborate field study may be rendered useless if population levels are not sufficiently high to allow effects to manifest themselves or be detected using mark–recapture methods. An investigator may have to search for proper environmental conditions or wait for suitable population levels before performing an experiment. Again, preliminary surveys of animal abundance may be helpful in designing the experiment and locating favorable test conditions.

Analysis of Population Data

Lapin (1975) states that to understand inductive statistics, one must first understand deductive statistics. Deduction is the process of drawing inferences about a particular set of observations from knowledge extracted about properties of the entire collection (i.e., universe). For example, if one is sampling from a normal distribution with mean μ and variance σ^2, then on the basis of the properties of a normal distribution, we know that the sample mean ($\bar{x}$) is also normally distributed with mean μ and with variance σ^2/n, where n is the size of the sample. Similarly, the development of a statistical test starts with a set of assumptions. From this set of assumptions, the probability distribution of a statistical test is developed through the process of deductive reasoning. Among the properties that can be deduced is the probability of observing a particular value of a test statistic, given that our initial hypothesis is correct. From these probabilities, the chance of obtaining a particular sample outcome can be computed.

The process of drawing conclusions from sample observations is the next step in hypothesis testing and is an extension of inductive logic. Induction is the process of drawing inferences about a universe from sample observations. If a particular sample results in a value of a test statistic that has a low probability of occurring, one could induce that the sample was not taken from the hypothesized distribution. As such, we would be inclined to reject the hypothesis. Because the conclusion is deduced from assumptions about an infinite population, a finite sample can never serve to prove the statistical hypothesis. Rather, sample observations can only be used to reject the hypothesis with a known chance of error.

Deductive and inductive reasoning are an integral part of the planning process in the design of experiments. The data analysis for an experiment should be planned at the same time the experiment is being designed and the objectives of the research determined. An investigator needs to envision, at least conceptually, an actual data set and the statistical analysis at the time the experiment is being planned (Allen and Cady 1982). In this way, the assumptions of the analysis can be checked against experimental procedures and anticipated results checked against various objectives of the field study. Often, inconsistencies can be found by this approach and corrected before serious consequences have occurred. Furthermore, the most efficient approach to an experiment only can be identified once the hypotheses are explicitly stated in statistical terms and the statistical analysis has been specified.

Federer (1973) gives the following characteristics of a useful research hypothesis: (1) descriptive of the phenomenon under investigation, (2) useful in predicting unknown facts, (3) clear, (4) testable, and (5) simple. For a research hypothesis to yield a useful statistical hypothesis, the phenomenon must also be described in terms of an estimable function of the parameters (Searle 1971, p. 189). The null hypothesis generally assumes that the effects being tested are nonexistent (e.g., H_0: $\mu_C - \mu_T = 0$). The alternative hypothesis then encompasses all other possible relationships. Hence, the alternative hypothesis is the contrapositive of the null hypothesis and must be true if the null hypothesis is false. In the case of one-tailed hypotheses, the null hypothesis is called a "composite" hypothesis because it possesses a range of values in addition to the proposition of equality (e.g., H_0: $\mu_C \geq \mu_T$). The alternative one-tailed hypothesis is also a composite hypothesis that is the contrapositive of the null hypothesis (e.g., H_a: $\mu_C < \mu_T$). As such, a set of testable hypotheses must be mutually exclusive and exhaustive. In some statistical tests, an indifference region may exist where neither the null nor the alternative hypothesis will be asserted if sample observations are intermediate in character.

If one could characterize a statistical population without error, then there would be no difficulty in deciding whether the null hypothesis was correct. However, acceptance or rejection of any hypothesis based on sample data is always subject to error. A statistical test is a decision rule that rejects the null hypothesis (H_0) when it is true no more often than $\alpha \cdot 100\%$ of the time, while rejecting a false H_0 with probability $1 - \beta > \alpha$. The probability of committing a Type I error is labeled α (i.e., the significance level of the test); and the probability of a Type II error, by β. The rejection of the null hypothesis when it is true is called a "Type I error" (Figure 1.4). The failure to reject the null hypothesis when it is false [the alternative hypothesis (H_a) is true] is called a "Type II error." The complement to the probability of a Type II error (β) is called the "power of the test" ($1 - \beta$). This power is the probability of accepting the alternative hypothesis when it is true. The statistical power of tests will be a focus of design considerations in later sections (Chapters 5 and 6).

The choice of an α-level should reflect the relative consequences of Type I and Type II errors in the experimental setting. For all else being equal, decreasing the α-level has the effect of increasing a Type II error. In laboratory experiments where the error variance can be relatively small, the traditional α-levels of 0.05 and 0.01 can be used while keeping the power of the test high. In population experiments where the sampling error and heterogeneity in animal abundance are likely to contribute to a large error variance, a low α-level may not be justified. If the consequences of Type I and Type II errors are approximately equivalent, an α-level of 0.10 or 0.20

	Accept H_o	Reject H_o
H_o True	Correct Decision $(1 - \alpha)$	Type I Error (α)
H_o False	Type II Error (β)	Correct Decision $(1 - \beta)$

Figure 1.4 Relationship between conclusions from a test of hypothesis and Type I (α) and Type II (β) error rates.

may be more appropriate in order to balance the two error rates. Otherwise, inordinate levels of field effort may be necessary at α-level of 0.05 and 0.01 for the power to be 0.95 or 0.99. Screening tests may deliberately establish large α-levels so that all advantageous treatments are detected. Confirmatory experiments may then set small α-levels in choosing the best treatment process.

Several characteristics of mark–recapture data to consider in the development of statistical tests of hypotheses include the following (Skalski 1985a):

1. Sampling errors of abundance estimates are not normally distributed for population levels and capture probabilities commonly encountered.
2. Sampling error of an abundance estimate is correlated with population size and may be unique at each replicate site.
3. Animal numbers are not normally distributed, and their distribution may change with plot size, spatial pattern of study plots, and treatment effects.
4. Treatment variances are likely to change with the mean abundance of populations.
5. Functional relationships between the mean response of populations and environmental and treatment effects are poorly understood.

An analysis of mark–recapture data must take into account these uncertainties and characteristics of the data to ensure valid inferences from a field experiment. However, the variable and uncertain nature of mark–recapture data makes the specification of exact tests impossible.

Inability to propose reasonable likelihood functions for multiplot, tag–recapture data has the effect of severely limiting the statistical theory available for construction of test statistics. Consequently, statistical tests must depend on the robustness of standard test distributions or nonparametric procedures. The approach taken in Chapters 5 and 6 will be to use parametric tests under various transformations and formulations of the data. Skalski (1985a) presents results on distributional behavior of nonparametric tests, likelihood ratio tests, and standard parametric procedures in analysis of mark–recapture data.

Bartlett (1947) lists four requirements for an ideal data transformation: (1) variance is unaffected by changes in mean response, (2) effects are additive and linear, (3) the arithmetic mean is an efficient estimate of the true mean, and (4) transformed random variable is normally distributed. A transformation that stabilizes the treatment variances, for example, may not necessarily produce effects that are additive and vice versa. Often, no single transformation will produce all the desired properties of a random variable. Eberhardt (1978) expressed the opinion that additivity under the proper response model is the most important consideration of a data transformation.

Most statistical methods for testing treatment effects rely on the two properties of additivity and homogeneity of variance. The latter refers to the situation in which the variance among replicates is the same under each treatment. This within-treatment variance will ordinarily consist of at least two distinct components. One source reflects random noise, as would exist in a homogeneous environment, and the other reflects plot-to-plot heterogeneity, which is uncontrolled and often results from unknown environmental factors that do have a real effect on response variables. Together, these two sources of variable produce the total, visible variance among replicate plots of a treatment.

The most direct approach to testing for treatment additivity across environments is an experimental approach corresponding to a generalized block experimental design. Figure 1.5 illustrates a situation where each block ($B = 3$) represents a different environmental state, where all treatments are present in each block, and each treatment ($T = 2$) is replicated ($R = 2$) within each block. The within-block (intrablock) variance of a treatment difference then provides a measure of noise variance as a baseline for testing whether the between-block (interblock) variance of this treatment difference exceeds the noise level. Such a test may be implemented through an analysis of variance (ANOVA) (Table 1.1).

The importance of additivity derives largely from subjective rather than statistical considerations, because it pertains to the subjective extrapolation of statistical inferences. If the B blocks in Figure 1.5 represent quite different environmental conditions, then a statistical confirmation that the difference between treatments shows no significant variation across these particu-

Block (Environment)	Randomized Block Design
1	T | C | T | C
2	C | T | T | C
3	T | T | C | C

Figure 1.5 Graphical representation of a randomized block experimental design with $T = 2$ treatments (control, C, and treatment, T) replicated twice ($R = 2$) in each of $B = 3$ blocks.

lar blocks lends credence to the subjective conclusion that this difference is invariant and will persist under other, untested environmental conditions.

"Additivity" as defined here represents only one potential type of *invariance relationship* in response to treatments, namely, where the treatment *differences in mean response* are invariant to change in background conditions. Multiplicativity represents an alternative type of invariance relationship that might exist between treatment responses, where ratios rather than differences are invariant to background changes. Such a model might be entertained if, for example, treatment effects are expected to be manifested through changes in the finite rate of population increase.

In population studies, the nature of environmental and treatment effects must be considered in the analysis of capture data. The assumption of additivity implies that a treatment effect will result in an increase or decrease in animal abundance by a fixed number of individuals independent of population size. For example, assumption of additivity would imply that each replicate population of a treatment would have on the average an increase of, say, 10 animals, regardless of the initial population size. Such an effect would have to be explainable in terms of the mechanisms of mortality, natality, and migration of a population.

Table 1.1

ANOVA Table for a Generalized Block Experimental Design Indicating Sources, Degrees of Freedom (df), Mean Squares (MS), and F-Test for Nonadditivity with a Two-Treatment (Control (C) vs. Treatment (T)) Design

Source	df	MS	F
Treatments (C vs. T)	$T - 1 = 1$	MST	
Blocks	$B - 1 = 2$	MSB	
Treatment × blocks	$(T - 1)(B - 1) = 2$	MSI	$F_{2,6} = \text{MSI/MSE}$
Replications within blocks	$(R - 1)BT = 6$	MSE	

Instead, a multiplicative effect, where a constant fractional change in abundance is anticipated (e.g., a 20% increase or decrease), seems more consistent with known mechanisms of population change. Hence, a multiplicative model and a logarithmic transformation of the data are generally suggested for analysis of population data.

Because the multiplicative model is a commonly considered alternative to the additive model, it is common practice to plot within-block treatment differences against block means or totals as a graphical method of exploratory data analysis. Tukey's 1-degree-of-freedom test for nonadditivity in a randomized block analysis of variance (Snedecor and Cochran 1980, pp. 283–285) provides a data-analytic test procedure that is especially sensitive against a multiplicative alternative to the additive model, and is becoming a routine data-analytic operation in many disciplines. The F-test for nonadditivity illustrated earlier (Table 1.1) is nonspecific in its sensitivity, as it is a general-purpose test against any alternative to the additive model.

Noting that a multiplicative relation among treatment responses on the original scale will become additive on the logarithmic scale, we see that this transformation of data that improves additivity also may improve homoscedasticity. Recall that the variance among replicate plots within a treatment consists of both random noise and causal variation due to plot-to-plot background variation in uncontrolled but causative environmental factors. On a strictly additive scale of measurement, each causative agent would induce exactly the same *additive change* in response in every treatment, even though each treatment has a different baseline of response; hence, the causal variance component of the total plot-to-plot variances within a treatment is expected to be the same for every treatment in a randomized experiment. Additivity and homoscedasticity are thus seen to be closely related issues, with additivity as a prerequisite to homoscedasticity.

The issue of homoscedasticity in the presence of additivity revolves around the question of whether, on the additive scale of measurement, the magnitude of the remaining noncausal components of noise variation in response are dependent on baseline levels of response, which do differ between treatments. These noncausal random variation components include pure random variation as well as any random sampling error that might be incurred through the mark–recapture process. In general, it is to be expected that heteroscedasticity will result from noncausal variation, thereby raising nuisance problems and creating a continued livelihood for statisticians.

2

Use of Preliminary
Survey Data

The probability that a treatment effect or a population change is detected in a wildlife investigation will depend on the size of change and relative magnitude of the error variance of the investigation. Consequently, to improve the performance of mark–recapture investigations, researchers need to consider the error sources that contribute to the overall variance of the observed change, and the manner in which design decisions may alter the magnitude of these error components. In the simplest circumstance, estimates of the error variance may be used to determine the level of field replication needed to detect a treatment effect under test conditions similar to those that provided the preliminary estimates. However, constantly changing environmental conditions and alternative field designs necessitate a more fundamental understanding of the effects of design decisions in the success of field investigations. Without such fundamental knowledge, the success of a field investigation is largely a matter of chance, and the researcher is relegated to the position of an interested bystander.

In this chapter, we explore the concept of variance component estimation and identify functional relationships that may exist between design parameters and changes in the anticipated magnitude of variance components. Relationships between variance components and design parameters such as plot size, trapping effort, and field replication will then be used in conjunction with cost functions to determine optimal field designs for wildlife investigations. Because the formulation of a cost function also is dependent on the availability of preliminary survey data, a short discussion of cost functions is presented as well. Additional material on cost functions for mark–recapture research can be found in Skalski (1985b).

Estimation of Variance Components

Wild populations are characterized by fluctuations in animal numbers through time as well as across the landscape. When animal abundance is estimated $(\hat{N})$ rather than enumerated (N), a third potential source of variability is the sampling error about the estimate. Which of these error sources will influence the performance of a field investigation will depend on the population comparisons tested or contrasts estimated. In abundance studies or comparative censuses, sampling error $[\text{Var}(\hat{N}_i \mid N_i)]$ is the only error source influencing the precision of the investigations. In contrast, in tests of treatment effects on animal abundance, the experimental error will consist of sampling error as well as spatial variability in abundance between replicate plots called the "plot-to-plot variance" $[\sigma_N^2]$.

In assessment and impact studies, the general structure of the error variance cannot be predicted without specifying the exact contrast to be tested. Some tests will be based on spatial variance in animal abundance, others based on temporal variation, and still others on a combination of temporal and spatial variance components. The only certainty is that the sampling error associated with estimation of abundance $(\hat{N})$ will always be a component in the overall error variance. Tests of impact based on sampling of environmental gradients or a series of independent point sources generally will be based on the plot-to-plot variance in abundance along with sampling error. For an impact assessment design presented in Chapter 6, temporal variances in abundance are used in the error term rather than spatial variability. The different structures for the error term in statistical tests of impact emphasize the interrelationship between hypothesis, field study design, and statistical analysis. Consequently, in performing a preliminary survey, a preconceived notion of the eventual hypotheses and analyses is necessary to ensure collection of proper data for subsequent sample size calculations.

The basic statistical formula underlying variance component analysis of multiplot animal abundance $(\hat{N})$ studies is the identity

$$\hat{N}_i = E(\hat{N}_i) + \{E(\hat{N}_i \mid N_i) - E(\hat{N}_i)\} + \{\hat{N}_i - E(\hat{N}_i \mid N_i)\}$$

where N_i is the actual abundance at site i and $\hat{N}_i$ is an estimate of that abundance based on the analysis of mark–recapture data. From this identity, the general total variance formula can be derived (Appendix 1) expressed in terms of conditional means and variances of $\hat{N}_i$ given the random variable N_i, which varies from plot to plot:

$$\text{Var}(\hat{N}_i) = \text{Var}(E\{\hat{N}_i \mid N_i\}) + E\{\text{Var}(\hat{N}_i \mid N_i)\}. \qquad (2.1)$$

A schematic illustration of what is meant by conditional means and variances is given in Figure 2.1, where the mean value of $\hat{N}_i$ at any value of N_i,

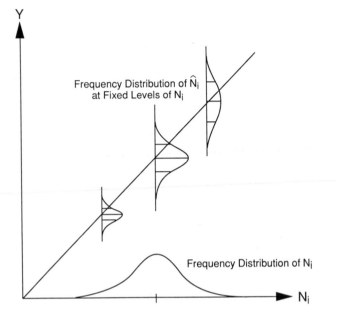

Figure 2.1 A schematic illustration of how the frequency distribution of $\hat{N}_i$ might change with N_i when $\hat{N}_i$ is an unbiased estimator for N_i.

$E(\hat{N}_i | N_i)$, is simply N_i. The shape of the distribution of $\hat{N}_i$ also depends on N_i in this illustration, and the variance $\mathrm{Var}(\hat{N}_i | N_i)$ is seen to increase with N_i.

In the special but relevant case where $E(\hat{N}_i | N_i) = N_i$ [i.e., where $\hat{N}_i$ is an unbiased estimator of animal abundance at site i, (N_i)], the general formula (2.1) reduces to

$$\mathrm{Var}(\hat{N}_i) = \sigma_N^2 + E\{\mathrm{Var}(\hat{N}_i | N_i)\}. \tag{2.2}$$

From Eq. (2.2), it is seen that the variance in population estimates among replicate sites is composed of two sources of variation. The variance among the $\hat{N}_i$ in a multiplot study exceeds the spatial variance in animal abundance (σ_N^2) by an amount representing the average sampling error of the estimates [i.e., $\overline{\mathrm{Var}(\hat{N}_i | N_i)}$]. Equation (2.2) is applicable for any unbiased abundance estimation scheme, and asymptotically correct for all maximum likelihood estimators of N. For this reason, Eq. (2.2) will be used throughout this book.

For most procedures, the sampling variance of the estimator $\hat{N}$ is directly proportional to the actual but unobserved abundance (N) on the plot, i.e.,

$$\text{Var}(\hat{N}_i \,|\, N_i) = N_i C_i. \qquad (2.3)$$

The proportionality factor C_i will depend on both sampling effort and sampling efficiency on the plot, and may therefore be specific to plot conditions. Consequently, the sampling error for an abundance estimate may be correlated with N_i for $i = 1, \ldots, l$.

Whether the proportionality factor, C_i, is homogeneous or heterogeneous among plots is a critical issue in the planning of a proposed experiment. It is imperative in designing the preliminary survey to build in the capability of the sampling program the ability of testing homogeneity of C_i values for any contemplated sampling and estimation procedure. Such capability also will enable the estimation of the sampling variance $[\text{Var}(\hat{N}_i \,|\, N_i)]$ for each plot sampled in the pilot study, so that in the event of heterogeneity among the $\hat{C}_i$ values, an estimate of the average sampling error variance $E\{\text{Var}(\hat{N}_i \,|\, N_i)\}$ can still be constructed in the form

$$\overline{\text{Var}(\hat{N}_i \,|\, N_i)} = \frac{1}{l}\sum_{i=1}^{l} \hat{\text{Var}}(\hat{N}_i \,|\, N_i) = \frac{1}{l}\sum_{i=1}^{l} \hat{N}_i \hat{C}_i. \qquad (2.4)$$

Inspection of Eq. (2.2) indicates that the empirical variance among the abundance estimates ($\hat{N}$ values) at the l plots

$$s_{\hat{N}}^2 = \frac{1}{l-1}\left[\sum_{i=1}^{l} \hat{N}_i^2 - \frac{1}{l}\left(\sum_{i=1}^{l} \hat{N}_i\right)^2\right] \qquad (2.5)$$

will tend to overestimate the corresponding variance among the actual but unobserved abundance values (N values) on these l plots

$$s_N^2 = \frac{1}{l-1}\left[\sum_{i=1}^{l} N_i^2 - \frac{1}{l}\left(\sum_{i=1}^{l} N_i\right)^2\right]$$

by an amount

$$\frac{1}{l}\sum_{i=1}^{l} \text{Var}(\hat{N}_i \,|\, N_i) = E(s_{\hat{N}}^2 - s_N^2 \,|\, N_1, \ldots, N_l)$$

which, in turn, is estimated by Eq. (2.4). An estimate of the unobservable s_N^2 is therefore available in the form

$$\hat{s}_N^2 = s_{\hat{N}}^2 - \overline{\text{Var}(\hat{N}_i \,|\, N_i)} \qquad (2.6)$$

and this, in turn, estimates the real target of the preliminary survey, the plot-to-plot variance of animal abundance, $\text{Var}(N_i) = \sigma_N^2$.

In the case where capture probabilities are shown to be homogeneous (i.e., where $C_i = C$ for $i = 1, \ldots, l$), the plot-to-plot variance component also can be estimated from the number of distinct animals trapped per plot, say, r_i, $i = 1, \ldots, l$. Let P denote the overall probability an animal is caught during the course of a mark–recapture study, where

$$E(r_i \mid N_i) = N_i P.$$

Writing the identity

$$r_i = E(r_i) + [E(r_i \mid N_i) - E(r_i)] + [r_i - E(r_i \mid N_i)]$$

and taking expected values as in Appendix 1, and letting $\mu_N = E(N)$ denote mean areal abundance, we obtain

$$r_i = \mu_N P + (N_i P - \mu_N P) + (r_i - N_i P). \tag{2.7}$$

Thus, the catch index r_i can be described by a linear model composed of a mean catch $\mu_N P$, deviations attributable to between-plot differences in abundance $(N_i P - \mu_N P)$, and sampling error $(r_i - N_i P)$. Taking the variance of (2.7), we find the variance of r_i to be

$$\mathrm{Var}(r_i) = P^2 \sigma_N^2 + \mu_N P (1 - P) \tag{2.8}$$

where $\mu_N P (1 - P)$ is the average sampling variance associated with the r_i and analogous to expression (2.4). Substituting the empirical estimators (the symbol $\triangleq$ is read as "estimates")

$$s_{r_i}^2 = \frac{1}{l - 1} \left[\sum_{i=1}^{l} r_i^2 - \frac{1}{l} \left(\sum_{i=1}^{l} r_i \right)^2 \right] \triangleq \mathrm{Var}(r_i),$$

$$\bar{r} = \frac{1}{l} \sum_{i=1}^{l} r_i \triangleq \mu_N P$$

for expressions in (2.8), the plot-to-plot variance estimate can be obtained from

$$s_{r_i}^2 = \hat{P}^2 \hat{\sigma}_N^2 + \bar{r}(1 - \hat{P}) \tag{2.9}$$

provided a suitable estimate of P is available. Typically, this implies that mark–recapture data conforming to a survey model would have been collected, allowing for the estimation of P. Although index data from a preliminary survey can be used in variance component analysis, a survey model is needed to provide a data structure necessary to extract the parameters of interest (i.e., σ_N^2 and μ_N). The advantage Eq. (2.9) provides is the ability to pool capture data across plots when capture numbers are low, but capture probabilities are homogeneous, to yield a precise estimate of σ_N^2.

EXAMPLE To illustrate the estimation of the plot-to-plot variance component, removal–trapping data of deer mice (*Peromyscus maniculatus*) from six 1-ha study plots located in the Peceance Basin, Colorado (Skalski et al. 1982a) will be used (Table 2.1). Using the abundance estimates derived from the removal model (Zippin 1956, 1958), Eq. (2.6) provides the variance component estimate

$$\hat{\sigma}_N^2 = 196.510 - 9.017 = 187.493$$

Table 2.1

Daily Capture Data (n_{ij}), Abundance Estimates ($\hat{N}_i$), and Variance Estimates
$[(\text{Vâr}(\hat{N}_i \mid N_i)]$ for *Peromyscus maniculatus* from Six 1-ha Study Plots

	Plot						
	1	2	3	4	5	6	Totals
n_{i1}	18	38	28	10	13	13	120
n_{i2}	12	9	6	8	6	5	46
n_{i3}	3	7	5	2	3	1	21
r_i	33	54	39	20	22	19	$\bar{r} = 31.167$
$\hat{N}_i$	36.982	56.853	40.832	23.434	24.655	19.641	$\bar{N}_i = 33.733$
$\text{Vâr}(\hat{N}_i \mid N_i)$	15.031	6.541	3.976	17.317	10.021	1.215	$\overline{\text{Vâr}(\hat{N}_i \mid N_i)} = 9.017$

while

$$\bar{\hat{N}}_i = 33.733$$

provides an estimate of mean animal abundance (μ_N). Alternatively, a test of homogeneous capture probabilities $[P(Q_5 \geq 2.29) = 0.808$, Skalski and Robson (1979)] indicates that capture data can be pooled across plots providing an estimated daily capture probability of $\hat{p} = 0.5920$ or a 3-day capture probability of

$$\hat{P} = [1 - (1 - \hat{p})^3] = 0.9321.$$

From Eq. (2.9), it then follows that the estimated plot-to-plot variance component based on the analysis of catch indices is

$$188.567 = (.9321)^2 \sigma_N^2 + 31.167(1 - 0.9321)$$

or

$$\hat{\sigma}_N^2 = \frac{188.567 - 31.167(1 - 0.9321)}{(0.9321)^2} = 214.605$$

and agreeing reasonably well with the previous estimate of 187.493. The same general approach can be used with other survey techniques and other variance structures to be encountered in this book.

Effect of Design Decisions on the Magnitude of Error Variances

To take full advantage of the information the preceding variance component analysis provides, an understanding of how changes in design parameters influence the magnitude of these error terms also is necessary. In this way, the performance of alternative field designs and allocations of sampling effort can be explored and the needed sampling effort under varying environmental conditions can be projected. For instance, sampling error [i.e.,

Var($\hat{N}_i | N_i$)] will be affected by changes in trapping effort (including trap density and duration of trapping) and population size, which, in turn, is a function of plot size and population density. In contrast, the plot-to-plot variance (σ_N^2) is influenced by plot size, plot dispersion in the landscape, and changes in animal density. These interrelated influences on the overall error variance need to be considered in the design of a field study whose purpose is to test hypotheses concerning the abundance of wild populations.

Influence of Trap Density

The proportionality factor C_i in Eq. (2.3) shows sampling error [i.e., Var($\hat{N}_i | N_i$)] to be a function of capture probabilities, which are, in turn, a function of trapping effort and the propensity of the animals to be caught. Consequently, the anticipated sampling error in a field study can be projected on the basis of the anticipated animal abundance and the per-period capture probabilities. Such projections are adequate for design purposes if the duration of the trapping effort can be adjusted until a predetermined fraction of the population has been caught. Practicality often dictates, however, that the duration of the sampling program be fixed (e.g., 5 or 10 days) and that trap density be adjusted to achieve a desired level of capture. Under such circumstances, the functional relationship between trap density and capture probabilities must be understood in order to project sampling error as a function of trap effort. Ideally, both trap density and trapping duration would be adjusted before the field experiment to optimally allocate sampling resources to minimize the error variance for a fixed research budget.

Insights into modeling of capture probability as a function of trap density can be gained by contemplating the relationship between survey travel time and sample size as presented by Jessen (1978). In the latter case, our attention focuses on the expected distance traveled in order to visit a given number of sample points located within a region of given size (Figure 2.2). In the case of trapping, we focus on a given animal whose territory constitutes the region in question and whose daily foray constitutes a path within this region (Figure 2.3). Assuming that the path has random-walk properties similar to a route followed by a survey enumerator in visiting a random sample of points (e.g., farms) in a political region, we may ask the inverse-sampling question; namely, given the distance traveled, what is the expected number of points (traps) visited by an animal wandering a trapping grid? Capture probability will be modeled as an increasing function of this expected number.

In the survey travel problem found in Jessen (1978) where territory size A and number of points visited f are fixed, the expected distance traveled $\bar{d}$

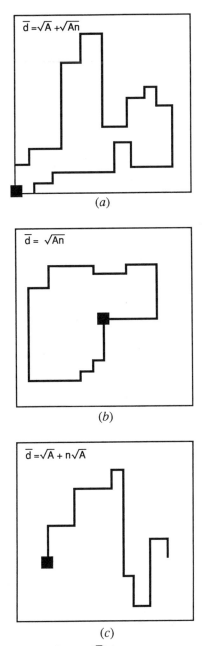

Figure 2.2 Expected travel distance ($\overline{d}$) for a survey enumerator driving on a road network to visit n random points. (a) Random arrangement, home in corner, $\overline{d} = \sqrt{A} + \sqrt{An}$; (b) random arrangement, home in center, $\overline{d} = \sqrt{An}$; (c) random arrangement, nonreversing path, $\overline{d} = \sqrt{A} + n\sqrt{A}$.

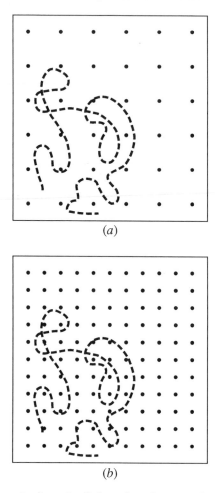

Figure 2.3 Schematic of an animal's foray through a serpentine array of traps of different densities (a, b). Probability of capture will be a function of trap density and the frequency of contact between the animal and the trapping device.

between the f randomly located points in A is given by

$$\bar{d} = f\left\{\sqrt{A}\left[\alpha + \beta\left(\frac{1}{\sqrt{f}}\right) + \gamma\left(\frac{1}{f}\right)\right]\right\}$$

where the coefficients α, β, and γ are dependent on the particular restrictions placed on the travel strategy. The bracketed term $\{\ \}$ represents the average travel distance between sample points. To relate this formula to the inverse sampling problem where travel distance d is fixed and the number of visited points is random, we may divide distance traveled by the average distance between stops in order to estimate the expected number of stops.

On the supposition that the animal's foray path might have random features analogous to the survey enumerator's shortest route through a random array of points, we thus suggest that the expected number of trap zones M visited on a daily foray might depend on trap density $\phi = f/A$ through the reciprocal relation

$$M = \frac{d}{\alpha\sqrt{A} + \beta\sqrt{\dfrac{A}{f}} + \gamma\dfrac{A}{f\sqrt{A}}} = \frac{1}{\alpha' + \beta'\left(\dfrac{1}{\sqrt{\phi}}\right) - \gamma'\left(\dfrac{1}{\phi}\right)} \qquad (2.10)$$

where f is the number of traps at the site. In the last equality, the animal's foray distance d and territory size A have been treated as constants and absorbed into the other constants of the function of trap density ϕ.

Visiting a trap zone exposes the animal to risk of capture, and assuming this risk to be constant and independent between successive traps, we can calculate the probability of capture on a foray that would otherwise have visited M trap zones. On the first encounter, there is some constant probability θ of being caught; with probability $1 - \theta$ the animal remains at large and proceeds to the next trap encounter, where, again, the probability of capture is θ. The probability of avoiding capture in the first two encounters is $(1 - \theta)^2$ under the independence assumption, and multiplying this by the probability of capture (θ) gives the probability of being captured in the third trap on the route. The combined probability of capture at some point on a path through M trap zones is therefore

$$p = \theta + (1 - \theta)\theta + (1 - \theta)^2\theta + \cdots + (1 - \theta)^{M-1}\theta$$
$$= 1 - (1 - \theta)^M \qquad (2.11)$$

which is simply one minus the probability of avoiding capture in all of the M encounters. Substituting the expression (2.10) into (2.11) then provides a model for capture probability p as a function of trap density ϕ,

$$p \approx 1 - (1 - \theta)^M = 1 - e^{+M\ln(1-\theta)}$$
$$= 1 - \exp\left[-\frac{1}{\alpha + \beta\left(\dfrac{1}{\sqrt{\phi}}\right) + \gamma\left(\dfrac{1}{\phi}\right)}\right] \qquad (2.12)$$

where, in the latter expression, the positive factor and constant $-\ln(1 - \theta)$ has been absorbed into the (also positive) coefficients of expression (2.10).

Note that the parameter setting $\alpha = \beta = 0$ and $\gamma = 1/c$ yields the conventional model (Seber 1982, p. 296)

$$p = 1 - e^{-\{1/[\gamma(1/\phi)]\}} = 1 - e^{-c\phi} = 1 - e^{-cf/A} \qquad (2.13)$$

with the "Poisson catchability coefficient" c. Model (2.12), transformable into a quadratic regression function of the reciprocal square root of trap density (ϕ),

$$\frac{1}{\ln\left(\dfrac{1}{1-p}\right)} = \alpha + \beta\left(\frac{1}{\sqrt{\phi}}\right) + \gamma\left(\frac{1}{\sqrt{\phi}}\right)^2$$

thus includes the conventional model (2.13) as a special case. The validity of the speculations leading to model (2.12) is therefore not a critical issue, but it is critical to note that in either the conventional capture probability model (2.13) or the more general model (2.12), the trapping effort parameter ϕ measures trap density rather than simply number of traps (f) as found in Seber (1982).

EXAMPLE   Trapping sessions in small mammal populations were conducted using replicate 1-ha plots in a bunchgrass–sagebrush (*Agropyron spicatum–Artemisia tridentata*) habitat to field test the appropriateness of the catch–effort models (2.12) and (2.13) in wildlife surveys (Skalski et al. 1983b). During May 1981, three plots serially located 0.5 km apart were used in the trial. During May 1982, the populations were retrapped on the original three plots plus an additional replicate plot located 0.5 km from the next-nearest site (Figure 2.4). Different levels (36, 64, 100, and 196 traps per hectare) of trap density were randomly assigned to the plots prior to each study. Removal trapping was conducted simultaneously at all plots for three consecutive nights, and numbers and species of animal captured were recorded. The purpose of the second trial and new randomization in 1982 was to ensure that any observed relationship between trapping effort and catch per unit (CPUE) was not the result of unrelated phenomena.

Totals of 142 and 52 mice [Great Basin pocket mouse (*Perognathus parvus*) constituted 63% (1981) and 87% (1982) of the catch; remaining individuals were deer mice (*P. maniculatus*) and northern grasshopper mice (*Onychomys leucogaster*)] were captured during the three nights of trapping in 1981 and 1982, respectively (Table 2.2). Weighted nonlinear least squares fitted to model (2.13) resulted in estimated Poisson catch coefficients ($\hat{c}$) of $0.0087[\hat{SE}(\hat{c}) = 0.00049]$ and $0.0128[\hat{SE}(\hat{c}) = 0.0043]$ for 1981 and 1982, respectively (Figure 2.5). Skalski et al. (1983b), using likelihood ratio tests, rejected a linear catch–effort model in favor of model (2.13) based on the data in Table 2.2.

The practical implication of models (2.12) and (2.13) is that although a greater catch (or equivalently, a greater capture probability) can be anticipated with higher trap densities, a diminishing return (i.e., lower CPUE) can be expected when increasing trap density (Figure 2.5). Model (2.13) may be used to project this relationship and its effect on sampling precision. On the basis of the catch–effort models, sampling error now can be expressed as a function of trap density, a design parameter available for manipulation. In subsequent chapters, model (2.13) will be used to illustrate

Trapping effort

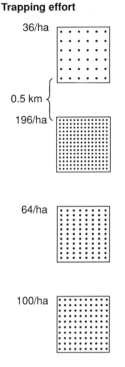

36/ha

0.5 km

196/ha

64/ha

100/ha

Figure 2.4 Arrangement of study plots and assignment of trapping effort in the small-mammal trials used in modeling catch–effort relationships.

how catch–effort relationships can be used to determine the optimal allocation of trapping effort (f_i) at a study plot. Skalski (1985a, pp. 32–50) illustrates how model (2.13) can be used to estimate the areal mean abundance (μ_N) of wild populations based on the field design of Figure 2.4.

Table 2.2

Trap Density ($f_i/$ha), Small Mammals Caught (n_i), and Observed CPUE (n_i/f_i) on 1-ha Plots in Southeast Washington

Replicate plot	May 1981			May 1982		
	Trap density ($f_i/$ha)	Catch (n_i)	CPUE (n_i/f_i)	Trap density ($f_i/$ha)	Catch (n_i)	CPUE (n_i/f_i)
1	36	26	0.722	196	20	0.102
2	196	77	0.393	36	9	0.250
3	64	39	0.609	100	14	0.140
4	—	—	—	64	9	0.140

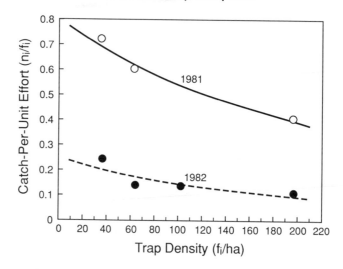

Figure 2.5　Relationship observed between the catch-per-unit effort and trap density exhibited by the 1981 and 1982 trap data and weighted nonlinear least-squares curves based on catch–effort model (2.13).

Influence of Animal Abundance

The magnitude of the plot-to-plot component (σ_N^2) of the experimental error variance will depend on the spatial variability in animal abundance, which, in turn, is often seen to be a function of mean animal abundance (μ_N). For any given plot size (A), the variance of N for many organisms is found to conform reasonably well to the negative binomial type of variance function

$$\sigma_N^2 = \mu_N + \frac{\mu_N^2}{R} \tag{2.14}$$

if not to the negative binomial frequency distribution. The latter distribution among plots could arise from a spatial distribution in which the number of animals (N) associated with a given plot is a Poisson random variable with parameter $A\lambda$, which is unique to that plot and that varies among plots according to a gamma frequency distribution. The local conditions at a plot are represented by λ, the expected animal density at that site. The coefficient $1/R$ in Eq. (2.14) then represents the squared coefficient of variation [i.e., $CV = \sqrt{1/R}$] of this gamma distribution on how λ varies between sites. Variance function (2.14) represents but a particular case again of the general variance component law (Appendix 1) rewritten as

$$\begin{aligned}
\text{Var}(N) &= E_\lambda[\text{Var}(N\,|\,\lambda)] + \text{Var}_\lambda[E(N\,|\,\lambda)] \\
&= E_\lambda[\text{Var}(N\,|\,\lambda)] + [E(N)]^2 \left\{ \frac{\text{Var}_\lambda[E(N\,|\,\lambda)]}{[E(N)]^2} \right\} \\
&= E_\lambda[\text{Var}(N\,|\,\lambda)] + [E(N)]^2\{CV_\lambda[E(N\,|\,\lambda)]\}^2.
\end{aligned}$$
(2.15)

A key feature leading from Eq. (2.15) back to (2.14) is the assumption of *local* randomness in the spatial distribution of animals resulting in N being locally Poisson-distributed at a site, with variance equal to mean; i.e., where

$$\text{Var}(N\,|\,\lambda) = E(N\,|\,\lambda) = A\lambda.$$
(2.16)

The first component μ_N in Eq. (2.14) is then simply the mean of the expected abundance $(A\lambda)$ over all local conditions

$$\mu_N = E_\lambda[E(N\,|\,\lambda)] = E_\lambda[A\lambda].$$

The second component in Eq. (2.14) is the plot-to-plot variance in expected abundance levels between sites $[E(N\,|\,\lambda) = A\lambda]$, but reexpressed in terms of the CV of $A\lambda$, as in Eq. (2.15). Under the key assumption of Poisson variability locally (2.16), the unconditional variance of N is thus expressible as

$$\begin{aligned}
\text{Var}(N) &= E(N) + [E(N)]^2\{CV[E(N\,|\,\lambda)]\}^2 \\
&= \mu_N + \mu_N^2\{CV(A\lambda)\}
\end{aligned}$$
(2.17)

without invoking any further assumptions. The variance function (2.14), which arises from a negative binomial model wherein the frequency distribution of λ across plots is a gamma distribution, now is also seen to hold irrespective of the distribution of local conditions λ, provided *local* randomness exists.

The reason for reexpressing the second component of (2.17) in terms of the CV of $E(N\,|\,\lambda)$ rather than the variance of $E(N\,|\,\lambda)$ is the expected invariance of the CV under different levels of overall abundance (μ_N). For instance, Eberhardt (1978) observed this invariance among CV values associated with population indices. Within limits, reduction in mean abundance (μ_N) should have merely a dilution effect. The result should be a common scalar reduction in all plot-specific Poisson density parameters λ, and reducing the plot-to-plot variance of λ by the square of this scalar, and thus having no effect on the CV of λ. Wide ecologic applicability of variance function (2.14) is not surprising in light of such considerations, and is not dependent on the negative binomial model.

The mean-to-variance relationships observed in model formulations (2.14, 2.17) have two practical implications for wildlife field experiments. First, the plot-to-plot variance (σ_N^2) can be estimated given no prior survey data to be at least as great as the population abundance expected on the replicate plots (i.e., $\sigma_N^2 \geq \mu_N$). The discussions below will suggest possibly $\sigma_N^2 \approx \mu_N$ when plots are in close proximity while $\sigma_N^2 > \mu_N$ when plots are several to many home ranges apart. Second, animal abundance can be expected to influence both sampling error and between-plot variance. Variance formula (2.3) suggests that the coefficient of variation (CV) for sampling error in an abundance study decreases as abundance increases (i.e., $CV \to 0$ as $N \to \infty$),

$$CV(\hat{N}|N) = \sqrt{\frac{C}{N}}. \tag{2.18}$$

Similarly, Eq. (2.14) as well as empirical data (Figure 2.6) suggest that the CV associated with plot-to-plot variance component (σ_N^2) also decreases as the expected abundance (μ_N) increases, implying

$$\frac{\sigma_N}{\mu_N} = \sqrt{\frac{1}{\mu_N} + \frac{1}{R}}. \tag{2.19}$$

Consequently, a change in mean abundance will have a joint effect on the variance components associated with the precision of a field experiment, expressed by Eq. (2.2). Greater animal abundance will generally improve the

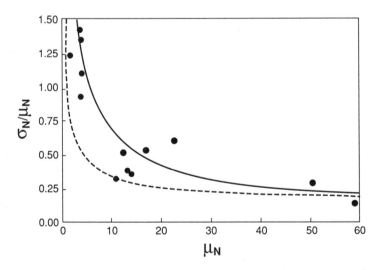

Figure 2.6 Observed relationship between the plot-to-plot variance component (expressed here as a $CV = \sigma_N/\mu_N$) and the mean abundance (μ_N) of small-mammal populations on 1-ha study plots. Data sets are from nine species in two western states with curves based on the negative binomial-type variance formula [(2.20), dashed line] and Taylor's power law [(2.22), solid line].

precision or power of field studies. Alternatively, the phenomenon exhibited by (2.18) in concert with (2.19) can make the detection of treatment effects highly unlikely in circumstances of low animal abundance. Researchers faced with performing a field experiment may need to delay implementation until environmental conditions and abundance levels are favorable.

EXAMPLE   Data from small mammal populations seem to agree with the negative-binomial type of variance function (2.14). In 13 data sets from multiplot field studies on small mammal species analyzed by Skalski (1985a), the size of the estimated plot-to-plot variance ($\hat{\sigma}_N^2$) always exceeded the estimate of mean abundance ($\hat{\mu}_N$). The data represented mark–recapture surveys on nine different species (*Dipodomy ordii, Eutamius minimus, Microtus montanus, Onychomys leucogaster, Perognathus flavus, Perognathus parvus, Peromyscus maniculatus, Reithrodontomys* spp., *Spermophylus tridecimlineatus*) in grass or shrub-steppe habitats of Washington and Colorado. All field studies used 1-ha study plots with as many as eight replicate plots in a single habitat type. Least squares fit of Eq. (2.14) to the 13 estimates of the plot-to-plot variance and mean abundance produced the relationship

$$\sigma_N^2 = \mu_N + 0.0315\mu_N^2 \qquad (2.20)$$

which agrees reasonably well with the field data (Figure 2.6).

In agricultural and ecological studies, a common alternative to the negative binomial variance formula is Taylor's power law (Taylor 1961) where

$$\sigma^2 = \alpha\mu^\beta. \qquad (2.21)$$

Using nonlinear least squares and the data in Figure 2.6, Taylor's power law projects the relationship

$$\sigma_N^2 = 11.4\mu_N^{0.63} \qquad (2.22)$$

for small-mammal abundance on 1-ha study plots. Again, this commonly used empirical formulation agrees well with the observed relationship between mean abundance and plot-to-plot variance of small mammal populations (Figure 2.6). The reason for the agreement is that the Taylor power law results in a family of curves similar to (2.14). An important distinction, however, is that Eqs. (2.21) and (2.22) predict that the $CV \to 0$ as $\mu_N \to \infty$ whereas the negative binomial variance law (2.19) predicts a lower bound of $1/\sqrt{R}$ for the CV of the plot-to-plot variance component. Previously, authors have considered R to be a characteristic of a species and invariant with respect to habitat.

Influence of Plot Size

The negative binomial type variance function (2.14) provided a model for the variance in animal abundance among plots of some fixed size when spatially dispersed in the study area to ensure independence and to sample the range of environmental conditions. The magnitude of this plot-to-plot variance was seen to depend on animal density in the study area in such a manner that the higher the density, the lower the coefficient of variation (CV) among plots of the given size, subject to the constraint that a minimum value for the CV did exist no matter how high the density. Both this limiting CV value and the rate at which it is approached depend on the size of the study plot. In the case of Poisson variation at very local levels within a study area [i.e., Eq. (2.14)], the relationship between plot CV and plot size will be found to be well approximated by simple empirical functions that can be useful in calculating an appropriate plot size.

A profoundly significant property of Poisson variation is the additivity of Poisson parameters; i.e., a sum of *independent* Poisson random variables is itself a Poisson random variable, and its parameter is simply the sum of the individual Poisson parameters. The local Poisson variation of N at a plot might thus be viewed in higher resolution as a sum of independent subplot chance variables with possibly different local density parameters. Adjacent subplots (which make up the plot) would be expected to have very similar parameters, but the degree of similarity should, on average, decrease with the distance separating subplots. In rows of subplots (Figure 2.7), the adjoining λ_i and λ_{i+1} may be highly correlated. Consequently, even though the random deviations $N_i - \lambda_i$ and $N_{i+1} - \lambda_{i+1}$ are independent Poisson deviates, the total variance (Appendix 1) of the sum $N_i + N_{i+1}$ becomes inflated by the high correlation (ρ_1) between the λ values of the adjacent subplots, where

$$\mathrm{Var}(N_i + N_{i+1}) = E[\lambda_i + \lambda_{i+1}] + \mathrm{Var}(\lambda_i + \lambda_{i+1})$$
$$= [E(\lambda_i) + E(\lambda_{i+1})] + [\mathrm{Var}(\lambda_i) + \mathrm{Var}(\lambda_{i+1}) + 2\mathrm{Cov}(\lambda_i, \lambda_{i+1})].$$

Interpretation of these variances and covariances is facilitated by considering a mosaic environment in which local heterogeneity exists but no trends occur. In the absence of any gradient in the study areas, all λ values have the same mean (μ_λ) and variance ($\sigma_\lambda^2 = \mu_\lambda^2[CV(\lambda)]^2$), giving the interpretation

$$\mathrm{Var}(N_i + N_{i+1}) = 2\mu_\lambda + \mu_\lambda^2[CV(\lambda)]^2[2 + 2\rho_1].$$

Similarly,

$$\mathrm{Var}(N_i + N_{i+1} + N_{i+2}) = 3\mu_\lambda + \mu_\lambda^2[CV(\lambda)]^2[3 + 2(2\rho_1 + \rho_2)]$$

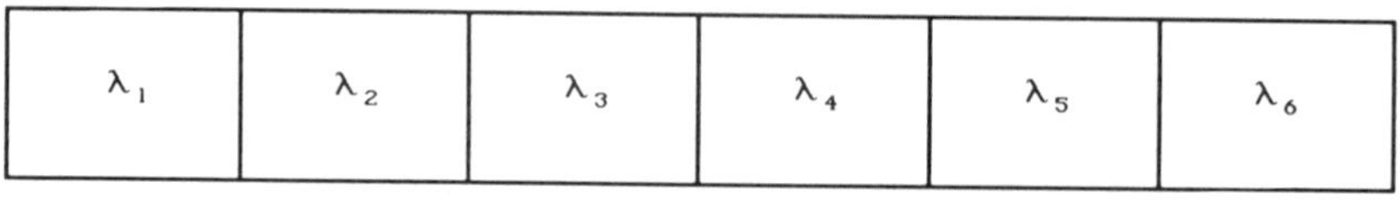

λ_1	λ_2	λ_3	λ_4	λ_5	λ_6

Figure 2.7 Conceptualization of a study area being composed of adjacent subplots, each with local Poisson variation (λ_i).

and

$$\text{Var}(N_i + N_{i+1} + N_{i+2} + N_{i+3}) = 4\mu_\lambda + \mu_\lambda^2[CV(\lambda)]^2[4 + 2(3\rho_1 + 2\rho_2 + \rho_3)]$$
(2.23)

where ρ_d, the correlation between λ_i and λ_{i+d}, is a decreasing function of d, $\rho_1 > \rho_2 > \rho_3 > \ldots > 0$. In ecological applications, this decreasing autocorrelation function is often found to exhibit the exponential decay pattern of a first-order autoregressive model, where $\rho_d = \rho^d$. The variance in abundance for plots composed of m segments spatially dispersed over the study area would then take the form

$$\text{Var}(N_i + N_{i+1} + \cdots + N_{i+m})$$

$$= m\mu_\lambda + \frac{\mu_\lambda^2[CV(\lambda)]^2\left\{m(1 + \rho) - \dfrac{2\rho}{1 - \rho}(1 - \rho^m)\right\}}{(1 - \rho)}.$$
(2.24)

A long strip plot of area A, envisioned as a linear array of subplots, each of unit area, generating a total Poisson count $N = N_1 + N_2 + \ldots + N_A$ with a local mean $\lambda_1 + \lambda_2 + \ldots + \lambda_A$, might thus be expected to show a global mean $\mu_N = A\mu_\lambda$ when averaged over all possible plots at a site, and a plot-to-plot variance of the form (2.17) or possibly (2.24).

A key point to note in these variance relationships is that the coefficient of $\mu_\lambda^2[CV(\lambda)]^2$ is a constant that lies between the limits A and A^2 (or m and m^2, since $A = m \geq 1$ in this case). If adjacent subplots exhibited no autocorrelation between λ_i and λ_{i+1}, then $\rho = 0$ and the coefficient in question is seen to equal $A = m$ in both (2.23) and (2.24). If λ_i and λ_{i+1} were perfectly correlated ($\lambda_i = \lambda$ for $i = 1, \ldots, A$) so that $\rho_d = 1$ for all d, then it is readily seen in (2.23), and not so readily in (2.24), the coefficient of $\mu_\lambda^2[CV(\lambda)]^2$ becomes $A^2 = m^2$.

These limits, A and A^2, remain valid not only for any autocorrelation function (ρ_d) but also for any plot shape, provided the variance function (2.14) of mean abundance holds, implying local Poisson variation. The implication of this simple result

$$\frac{CV(\lambda)}{\sqrt{m}} \leq CV\left(\sum_i^m \lambda_i\right) \leq CV(\lambda)$$

is that when variance relationship (2.14) holds, we should find that plot-to-plot variance of density (N/A) should conform to some function bounded by

$$\frac{D}{A} + \frac{D^2\omega}{A} < \text{Var}\left(\frac{N}{A}\right) < \frac{D}{A} + D^2\omega$$

where $D = \mu_\lambda$ denotes mean density at the site and $\sqrt{\omega} = CV(\lambda)$ denotes the coefficient of variation of λ at some arbitrarily small subplot size. (Note that A is measured here as $m \geq 1$, multiples of this subplot size, and D is mean density per subplot.)

A negative binomial distribution for animal abundance among plots would be obtained if all adjoining subplots within a plot of size A had a common density parameter λ, according to a gamma frequency distribution, in which case

$$\text{Var}\left(\frac{N}{A}\right) = \frac{D}{A} + D^2\omega.$$

In general, however, the choice of a particular function $g(A)$ to use in modeling plot-to-plot variance

$$\text{Var}\left(\frac{N}{A}\right) = \frac{D}{A} + D^2\omega g(A)$$

where

$$\frac{1}{A} \leq g(A) \leq 1 \qquad \text{for} \qquad A > 1$$

is largely empirical. An obvious and commonly used choice is the power function

$$g(A) = \frac{1}{A^b}, \qquad \text{where} \qquad 0 \leq b \leq 1$$

that is parsimonious in parameters, requiring empirical determination of the unknowns ω and b as well as the mean density D in

$$\text{Var}\left(\frac{N}{A}\right) = \frac{D}{A} + \frac{D^2\omega}{A^b}. \tag{2.25}$$

At very low densities, the second term becomes negligible and the Poisson variance formula applies. The coefficient of D/A in the first term of (2.25) is unity because of the assumption of uncorrelated Poisson deviates $(N_i - \lambda_i)$ among the A hypothetical subplots within each plot. Without this assumption, the unknown coefficient of D/A would become yet another parameter (η) to determine in (2.25) and resulting in the still more general formula

$$\text{Var}\left(\frac{N}{A}\right) = \frac{D\eta}{A} + \frac{D^2\omega}{A^b}. \tag{2.26}$$

At high densities, the second term dominates, and Eq. (2.25) reduces to H. Fairfield Smith's empirical variance law in relating the variance in experimental crop yield (Y) to plot size (A) in agronomic trials (Smith 1938), where

$$\text{Var}\!\left(\frac{Y}{A}\right) = \frac{\sigma^2}{A^b} \tag{2.27}$$

and where σ^2 is the variance among plots of unit area.

EXAMPLE Data requirements for estimating the variance function (2.25) are substantial, and in the case of small mammal experiments, a gradual accumulation of relevant data is the expected mode of operation. Other biological fields involving plants and lower animals are more fortunate in this regard, and we provide an illustrative example drawn from economic entomology. Beal (1939) enumerated Colorado potato beetles in each 2 ft of row in a potato field of 48 rows, each 96 ft long. Synthetic "plots" of increasing size were formed by overlaying successively coarser grids on this two-way data table. A summary of the data as presented by Jessen (1978) is found in Table 2.3. A mean beetle count per 2-ft segment provides a density estimate of $\hat{D} = 4.737$. Regression of the dependent variable

$$\ln\!\left(\frac{\text{Var}\!\left(\dfrac{N}{A}\right) - \dfrac{\hat{D}}{A}}{\hat{D}^2}\right)$$

on plot size (A) gives $\hat{b} = 0.241$ and $\hat{\omega} = 0.422$ in Eq. (2.25) with the resulting model

$$\text{Var}\!\left(\frac{N}{A}\right) = \frac{4.737}{A} + \frac{9.469}{A^{0.241}}.$$

Alternatively, deleting the Poisson variance term and fitting the H. Fairfield Smith's variance law (2.27) gives $\sigma^2 = 10.07$ and, again, $\hat{b} = 0.241$. Comparison of observed variances and those predicted by Eqs. (2.25) and (2.27) indicates that the Poisson component is clearly needed, but its effect quickly becomes negligible with increasing plot size (Table 2.4).

Earlier in the season, on the other hand, when D was small, the Poisson term would have dominated over a much broader range of plot sizes. Failure to include this term can result in a grossly misleading interpretation of variance behavior at low density. This is particularly evident in terms of the coefficient of variation; from Eq. (2.25) we have

$$CV(N) = \sqrt{\frac{1}{AD} + \frac{\omega}{A^b}}$$

Table 2.3

Variance in Colorado Potato Beetle Density (Beal 1939)
as a Function of Plot Dimension and Size

Plot dimension (ft)	Plot area (A) (ft^2)	$\hat{\text{Var}}(N/A)$
1×1	1	15.00
2×2	4	7.23
4×4	16	4.94
6×6	36	4.48
8×8	64	3.54

Table 2.4

Comparison of Observed Variance in Beetle Density (Beal 1939) and
That Predicted by Eqs. (2.25) and (2.27) as a Function of Plot Size (A)

Plot size (A)	Observed (Beal 1939)	$\dfrac{\hat{D}}{A} + \dfrac{\hat{D}_\omega^2}{A^b}$	$\dfrac{\hat{\sigma}^2}{A^b}$
1	15.00	14.21	10.07
4	7.23	7.97	7.18
16	4.94	5.15	5.14
36	4.48	4.13	4.22
64	3.54	3.55	3.68

which, when regarded as a function of D for a fixed plot size, is seen to increase without bound as $D \to 0$, due to the presence of the Poisson component. The H. Fairfield Smith's variance law, lacking a term corresponding to this Poisson component, may be inadequate and misleading in some circumstances where it has been applied (Robson 1974). Adequate representation of the variance function to capture its critical features in terms of A and D is a major key to informed planning of future experiments.

Influence of Plot Dispersal

The preceding development is based on the proximity principle that neighboring plots are more similar than distant plots and hence are expected to exhibit more similar levels of animal abundance. Expressed in terms of the "local Poisson" model, this principle implies that plots located close together must have similar Poisson density parameter values (due to their high autocorrelation) and hence, are expected to exhibit approximately Poisson variability between plots. In these circumstances, plot-to-plot variance should be approximately equal to the mean abundance per plot. Conversely, distant plots are expected to have dissimilar density parameters and to exhibit plot-to-plot variance of the negative binomial type, exceeding the mean abundance per plot by a term of the order of the square of mean abundance.

Design implications of this principle are that in a blocked experimental design for testing treatment effects, the plots within a block should be in close proximity while the blocks themselves should be well dispersed. Dispersing the blocks allows tests of the treatment effects over a range of abundance levels while building sufficient degrees of freedom to achieve adequate power for detecting a treatment effect.

EXAMPLE   Data from a preliminary sampling program conducted at Rocky Mountain Arsenal, Commerce City, Colorado, during Spring 1982 (Thomas et al. 1983) will be used to illustrate the proximity principle and the value of blocked designs. Eight 1-ha study plots in sand dropseed (*Sporobolus cryptandrus*) habitat were selected in order to estimate small-animal abundance (Figure 2.8). Within each of four areas, a pair of 1-ha plots was established for a total of $2l = 8$ plots. Within blocks, the distance between plots was less than 1 km, while distances between blocks ranged from 3 to 8 km.

At each plot, Sherman live traps were set 11 m apart in a 9×9 square grid with trapping conducted simultaneously at all eight plots. Trapping was conducted for 5 consecutive days (June 17–21); 2 days of live trapping for marking the animals followed by 3 days of removal trapping. Small mammals captured during the marking period were toe-clipped and released at point of capture.

Abundance estimates for the deer mouse (*P. maniculatus*) at each plot (Table 2.5) based on the Lincoln indices (Seber 1982) indicate an average abundance of 13.38 individuals with an average sampling error of $\hat{\text{Var}}(\hat{N}_i | N_i) = 38.76$. Alternatively, total variance in actual abundance (N) (rather than estimated abundance, $\hat{N}$) among the eight plots may be regarded as consisting of two components, one due to between-block differences and the other, to within-block differences. Expressed in the form of Eq. (2.1), the variance in abundance can be written as

$$\text{Var}(N) = \text{Var}[E(N \,|\, \text{block})] + E[\text{Var}(N \,|\, \text{block})], \qquad (2.28)$$

representing the between-block ($\text{Var}[E(N \,|\, \text{block})]$) plus the average within-block ($E[\text{Var}(N \,|\, \text{block})]$) variance components. Since only the estimated ($\hat{N}$) rather than the actual (N) abundance is available for each plot, these two components of the total variance of N are estimable only indirectly.

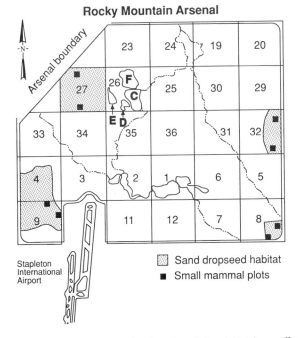

Figure 2.8 Map of study area showing location of the eight 1-ha small-mammal study plots distributed in a nested design.

Table 2.5

Lincoln Index Estimates of Deer Mice (*Peromyscus maniculatus*) Abundance on Four Pairs
of 1-ha Plots at Rocky Mountain Arsenal

	Sections								
	8		9		27		32		
	P_1	P_2	P_1	P_2	P_1	P_2	P_1	P_2	Average
Lincoln Index	18.25	10.00	16.50	16.60	10.67	29.00	2.00	4.00	13.38

The expression for the total variance of $\hat{N}$ is found by applying formula (2.1) twice, once
with regard to the effect of blocking and once with regard to the mark–recapture process used
in obtaining the estimates $(\hat{N})$, where

$$\begin{aligned}
\text{Var}(\hat{N}) &= \text{Var}[E(\hat{N}\,|\,\text{block})] + E[\text{Var}(\hat{N}\,|\,\text{block})] \\
&= \text{Var}[E[E(\hat{N}\,|\,N,\,\text{block})]] + E\{\text{Var}[E(\hat{N}\,|\,N,\,\text{block})] \\
&\quad + E[\text{Var}(\hat{N}\,|\,N,\,\text{block})]\} \\
&= \text{Var}[E(N\,|\,\text{block})] + E[\text{Var}(N\,|\,\text{block})] \\
&\quad + E[E[\text{Var}(\hat{N}\,|\,N)]]
\end{aligned} \tag{2.29}$$

since $\hat{N}$ is approximately unbiased [i.e., $E(\hat{N}\,|\,N) = N$]. The first two variance components of
$\hat{N}$ are the same as those of Eq. (2.28). The third term in (2.29) is the expected sampling error
$[\text{Var}(\hat{N}\,|\,N)]$ over blocks and plots within blocks.

A nested analysis of variance (Table 2.6) of $\hat{N}$ enables the estimation of the two variance
components of (2.28) by noting that the average observed sampling variance can be used to
estimate expected sampling error

$$\overline{\text{Var}(\hat{N}\,|\,N)} \triangleq E[E[\text{Var}(\hat{N}\,|\,N)]] = 38.76.$$

The within-block component of variance in abundance (N) is estimated from the plot-within-
blocks mean square (W) of Table 2.6, where

$$\hat{E}[\text{Var}(N\,|\,\text{block})] = 51.01 - 38.76 = 12.25$$

or more generally

$$W - \overline{\text{Var}(\hat{N}\,|\,N)} \triangleq E[\text{Var}(N\,|\,\text{block})]. \tag{2.30}$$

Table 2.6

One-Way Analysis of Variance of the Eight Abundance Estimates $(\hat{N})$ of *Peromyscus
maniculatus* from Survey at Rocky Mountain Arsenal

Source	df	MS	E(MS)
Blocks	$(l - 1) = (4 - 1)$ $= 3$	$B = 106.68$	$b \cdot \text{Var}[E(\hat{N}_i\,|\,\text{block})] + E[\text{Var}(\hat{N}_i\,|\,\text{block})]$ $+ E[E[\text{Var}(\hat{N}\,|\,N)]]$
Plots within blocks	$l(b - 1) = 4(2 - 1)$ $= 4$	$W = 51.01$	$E[\text{Var}(\hat{N}_i\,|\,\text{block})] + E[E[\text{Var}(\hat{N}\,|\,N)]]$

It remains to estimate the between-block variance component, which is computed from the blocks mean square (B) and the plots-within-blocks mean square (W), by the formula

$$\frac{B - W}{2} \doteq \text{Var}[E(\hat{N} \,|\, \text{block})] \tag{2.31}$$

giving the estimate

$$\hat{\text{Var}}[E(\hat{N} \,|\, \text{block})] = \frac{106.68 - 51.01}{2} = 27.84.$$

Having removed the noise due to mark–recapture sampling error, we thus arrive at estimates of the two components of (2.28):

$$\hat{\text{Var}}(N) = \hat{\text{Var}}[E(N \,|\, \text{block})] + \hat{E}\{\text{Var}(N \,|\, \text{block})\} \tag{2.32}$$

or more specifically

$$\hat{\text{Var}}(N) = 27.84 + 12.25 = 40.09.$$

The point of major interest, however, is that these estimates agree with the model of spatial variance; indeed, the within-block component estimate (12.25) stands in agreement with an assumption of Poisson variation at the local level defined by within-block plot distances. This variance component measures variation in local abundance and closely approximates the mean abundance estimate (13.38 animals per hectare), suggesting that Eq. (2.14) pertains with

$$\sigma_N^2 = \mu_N + \frac{\mu_N^2}{R} \doteq 13.38 + 27.84$$

and implying

$$\frac{\mu_N^2}{R} \doteq 27.84$$

or

$$\frac{1}{\hat{R}} = \frac{27.84}{(13.38)^2} = 0.156$$

[and a coefficient of variation for the Poisson parameters $\sqrt{0.156} = 0.395$, or $CV(\lambda) \doteq 39.5\%$]. For planning purposes, we might therefore adopt the variance function

$$\sigma_N^2 \approx \mu_N + 0.156\mu_N^2$$

for well-dispersed plots at this site while $\sigma^2 \approx \mu_N$ at the local level (≤ 1 km) of pairing or blocking to test treatment effect on this species. Similar effects of plot dispersal on the plot-to-plot variance were observed for other small mammal species and total species abundance at Rocky Mountain Arsenal (Thomas et al. 1983).

Estimation of Covariances and Correlations

For some field designs in Chapters 5 and 6, the principle of pairing is used to take advantage of a positive correlation in abundance between populations and reduce the overall error variance in test of effects. In some instances, study plots will be arranged so that control and treatment sites are paired in close proximity while similar site pairs are dispersed over the larger area. The variance of the difference in abundance estimates $(\hat{N})$ between control and treatment sites will, therefore, include a component for the spatial covariance between sites within pairs. Alternatively, field designs based on the analysis of estimates of abundance, both before and after application of a treatment, will include within the overall error variance a component for the temporal covariance in animal abundance. Consequently, preliminary survey data or historical records need to be analyzed to estimate these covariance components for subsequent sample size calculations.

When animal abundance at a series of study plots ($j = 1, 2, \ldots, l$) is estimated at time i and again at time $i + 1$, the temporal covariance in abundance can be estimated using the general covariance law

$$\mathrm{Cov}(\hat{N}_{ij}, \hat{N}_{i+1,j})$$
$$= \mathrm{Cov}[E(\hat{N}_{ij} \mid N_{ij}), E(\hat{N}_{i+1,j} \mid N_{i+1,j})] + E[\mathrm{Cov}(\hat{N}_{ij}, \hat{N}_{i+1,j} \mid N_{ij}, N_{i+1,j})].$$

Here a natural ordering exists between observations at time i and $i + 1$. In the special case where abundance estimators $(\hat{N})$ are approximately unbiased [i.e., $E(\hat{N} \mid N) = N$],

$$\mathrm{Cov}(\hat{N}_{ij}, \hat{N}_{i+1,j}) = \mathrm{Cov}(N_{ij}, N_{i+1,j}).$$

Consequently, the covariance in population abundance between time i and time $i+1$ can be estimated as

$$\mathrm{C\hat{o}v}(\hat{N}_{ij}, \hat{N}_{i+1,j}) \triangleq \mathrm{Cov}(N_{ij}, N_{i+1,j}).$$

The (product–moment) correlation between two random variables is expressed as a function of the covariance and variances of the random variables. The correlation in animal abundance between time i and $i + 1$ can be written as

$$\rho_{N_i, N_{i+1}} = \frac{\mathrm{Cov}(N_i, N_{i+1})}{\sqrt{\mathrm{Var}(N_i) \cdot \mathrm{Var}(N_{i+1})}}. \tag{2.33}$$

In a previous discussion, we noted that the variance of $\hat{N}$ includes in addition to plot-to-plot variability, sampling error. Consequently, an estimate of (2.33) based on the use of empirical variability among the estimates [i.e., $s_{N_i}^2 \triangleq \mathrm{Var}(N_i)$] will result in ρ being underestimated. As an alternative in population investigations, the temporal correlation should be estimated by

$$r_{N_i, N_{i+1}} = \frac{\mathrm{C\hat{o}v}(\hat{N}_i, \hat{N}_{i+1})}{\sqrt{\hat{s}_{N_i}^2 \cdot \hat{s}_{N_{i+1}}^2}}$$

where $\hat{s}_N^2$ is given by (2.6) to remove the sampling error present in abundance studies.

In preliminary studies where pairing or blocking is used (e.g., Figure 2.8), the sites within blocks do not have a natural ordering or class assignment on which the spatial covariance can be calculated. Instead, the intraclass correlation (Snedecor 1956, pp. 282–285) may be calculated to express the spatial correlation anticipated in paired (or blocked) experiments. A distinction between product–moment correlation and intraclass correlation is that in the latter case, there is homogeneity within each class in the sense that all elements are of equal rank and are interchangeable. In the case of Figure 2.8, for example, the two selected plots within a block have no unique qualities and would be exchangeable with any other two plots from that area that are adequately separated.

Intraclass correlation (ρ_I) is defined in this context as the ratio of the "between-blocks" component of variance, $\text{Var}[E(N \,|\, \text{block})]$, to the sum of the between- and within-block components

$$\rho_I = \frac{\text{Var}[E(N \,|\, \text{block})]}{\text{Var}[E(N \,|\, \text{block})] + E[\text{Var}(N \,|\, \text{block})]} = \frac{\text{Var}[E(N \,|\, \text{block})]}{\text{Var}(N)}$$

and provides a useful index for measuring the effectiveness of blocking. When ρ_I increases toward 1, the blocking approaches perfection, while at $\rho_I = 0$, the blocking is totally ineffective. As in the case of estimating product–moment correlation based on abundance estimates rather than actual abundance, the sampling error variance component due to estimation should be removed in the manner of (2.33) before estimating ρ_I. The numerator of ρ_I is estimated by formula (2.31) and the denominator by (2.32).

EXAMPLE For the two-stage field sampling illustrated in Figure 2.8, where between-block and within-block formulas were applied, giving 27.84 and 12.25, respectively, the intraclass correlaton is thus estimated to be $27.84/40.09 = 0.694$. Note that if the within-block variance component represents only Poisson variance, as appeared to be the case in this example, and if the between-block variance component is of the negative-binomial type (2.14), then ρ_I becomes an increasing function of mean abundance (μ_N)

$$\rho_I = \frac{\left(\mu_N + \dfrac{\mu_N^2}{R}\right)}{\mu_N + \left(\mu_N + \dfrac{\mu_N^2}{R}\right)} = \frac{1 + \dfrac{\mu_N}{R}}{2 + \dfrac{\mu_N}{R}} = \frac{\dfrac{1}{\mu_N} + \dfrac{1}{R}}{\dfrac{2}{\mu_N} + \dfrac{1}{R}},$$

which is bounded below by $\frac{1}{2}$ and increases toward 1 as μ_N increases. In the present example, $1/R$ was estimated to be 0.156, which would give

$$\rho_I = \frac{1 + 0.156\mu_N}{2 + 0.156\mu_N}.$$

At the estimated abundance of $\hat{\mu}_N = 13.38$, the intraclass correlation can be alternatively estimated at Rocky Mountain Arsenal to be

$$\rho_I = \frac{1 + 0.156(13.38)}{2 + 0.156(13.38)} = 0.755 \,.$$

This intraclass correlation could subsequently be used to determine the effects of pairing in field study designs at the site.

Cost Functions

Most design factors influencing the performance of a multiplot field study also influence the cost of conducting the research. Level of precision and research costs are functionally related through their common dependence on design factors determining the scope and magnitude of the research effort. Levels of such key factors as plot size, replication, trapping intensity, and duration can be associated in various combinations to produce alternative designs ranging from a few intensively studied plots to extensive designs with numerous plots scattered across the landscape that are only lightly sampled to estimate abundance. The statistical performance of an experiment may differ substantially among these alternatives, rendering some options unacceptable. Among study designs with acceptable statistical power, large differences in cost may exist. Statistical power is determined by the relationships between key design factors and the plot-to-plot variance and sampling error of abundance estimates; planning an efficient design also requires the quantification of cost dependencies on these key factors.

Constructing a cost function is a vital step in the planning stage of a field study to ensure that the design will fit within available research funds. Equally important, a cost function may be essential in identifying an optimal study design to provide the statistical performance required. Assembling the cost function requires identifying the various tasks in a field study and determining how the equipment and labor costs associated with these tasks enter the formulation. Costs may enter as (1) fixed expenses independent of the size of the study, (2) costs directly related to the number of study plots, and (3) costs that depend on trapping intensity and duration. Compiling a list of successive steps required to implement a field study may facilitate identification of cost components; for example, a study may consist of the following:

1. Research planning and preparation
2. Pilot study
3. Site selection
4. Site preparation and the application of "treatments"
5. Trapping
6. Site cleanup and departure
7. Data analysis and publication

Costs associated with these or similar steps can then be characterized as fixed expenses, costs dependent on the number of study plots, or trapping intensity. The latter two categories include travel costs that require separate consideration.

Travel Costs

Travel during a field study may be one of three types: travel to and from home or office and the research area, travel between plots, and intraplot travel. The first of these types of travel is the simplest to model. If H is the duration of the workday and t_0, the average travel time to the research area, then the length of the effective workday is reduced to $H - 2t_0$. Hence, travel to and from the research site has the effect of reducing the length of the workday and subsequently increasing labor requirements. Cost of travel within and between plots follows patterns similar to those of a traveling salesperson or social survey interviewer, and the costs associated with these patterns of travel have been extensively studied in other context. Jessen (1978, pp. 98–100) presents a detailed discussion of these costs in the context of socioeconomic surveys. Figure 2.2 illustrates some of these patterns with associated travel distance models.

Intraplot travel is readily modeled for an equally spaced grid of f traps in a rectangular plot of area A with a spacing of distance $\sqrt{A/f}$ between trap rows and columns if the outer rows and columns are located a half spacing in from the edge of the plot (Figure 2.9). The serpentine travel distance required to visit all f traps then includes the half spacing to the first trap in the corner, plus the $f - 1$ spacings between consecutive traps, plus half-spacing to exit the plot after the last trap is visited. This gives the total of f spacing distances, each of length $\sqrt{A/f}$ as total travel distance $f\sqrt{A/f} = \sqrt{Af}$.

Jessen (1978) also gives $\sqrt{An}$ as the formula for the expected total travel distance connecting n randomly located points in an area A starting from the center of the area. This same formula may therefore be applied to approximate the interplot travel distance once the worker has arrived at a study area of size A' containing a scatter of n plots. Travel costs will be proportional to these travel distances within (i.e., $\sqrt{Af}$) and between plots (i.e., $\sqrt{A'n}$).

Fixed Expenses

Funds available for actual implementation of the fieldwork are those funds remaining after subtracting fixed costs. This remainder, constituting the total operating funds to be apportioned among field costs, will be designated as C_0. Preoperational fixed expenses might include the cost of a pilot study, technical review of the resultant research plan, organization of data-entry

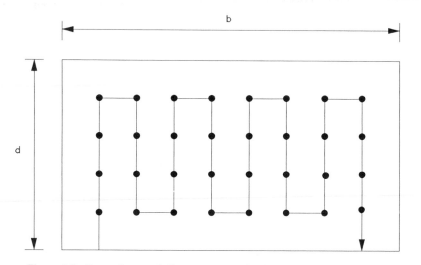

Figure 2.9 Serpentine travel distance over a grid of $f = n_r n_c$ equally spaced traps in a rectangular area $A = db$, $(d/n_r) = (b/n_c) = \sqrt{db/(n_r n_c)} = \sqrt{A/f}$.

systems, equipment selection and preparation, map reconnaissance, literature review, and preliminary computing of prospective statistical power curves. Such preliminary costs are largely independent of the size of the study and hence may be deducted before considering the other types of costs.

Costs Dependent on Number of Study Plots

Some components of cost are dependent on the size of the study only through the number and dispersal of study plots, being independent of the intensity of sampling within plots. These interplot expenses include such items as site selection costs and the interplot travel costs during the operational phase. Because they are independent of trapping intensity, these between-plot costs can be modeled separately for later merging with the within-plot costs to model the cost function.

The process of site selection may involve map selection of numerous areas to be initially visited and cursorily examined before final selection of study plots. If blocking is anticipated, this process might involve vegetation and soil surveys to provide a basis for partitioning plots into blocks to achieve within-block homogeneity. The site selection process is critical to the success and interpretation of the study, because all inferences are dependent on and restricted by this process which defines the population of inference. Documentation of the selection process, including information on the rejected as well as the accepted study plots, is advisable since the target population is defined by what it is not, as well as by what it is. Such documentation represents an additional expense, which should be incorporated into the cost model. In general, the total of those costs associated with site selection will be proportional to the number (n) of replicate plots per treatment, say, $C_1 n$.

Another significant interplot expense is the travel costs between sites, including the trapping phase, and at the initial and final phases of site preparation and cleanup, respectively. If sites are visited daily during a d-day trapping phase, then the combined interplot travel costs should be approximated by

$$C_2\sqrt{n} + C_3 d\sqrt{n}$$

where the $\sqrt{A}$ of Jessen's formula (Jessen 1978) has been absorbed into C_2 and C_3. Adding in the site selection costs then gives the total interplot expense

$$C_1 n + C_2\sqrt{n} + C_3 d\sqrt{n} \tag{2.34}$$

where the respective C-coefficients will depend on specific conditions of the study and may be estimated from the pilot study. Note that expression (2.34) consists of expenses from the following sources:

$$[\text{Site selection}] + \begin{bmatrix} \text{Interplot travel during} \\ \text{preparation and termination} \end{bmatrix} + \begin{bmatrix} \text{Interplot travel} \\ \text{during trapping} \end{bmatrix}.$$

Costs Dependent on Trapping Effort

Plot size (A) as well as number of traps per plot (f) enter the within-plot cost model. The initial expense of purchasing the traps is proportional to their number per treatment (nf). Cost of daily trap maintenance during the trapping phase is proportional to the number of trap days (ndf). Furthermore, the time and labor cost of handling captured animals is also proportional to ndf. To see this relationship, note that the number of animals caught per day at a plot is expected to be $\mu_N p$, where μ_N is mean abundance per plot of size A and p is the capture probability modeled by (2.13) as a function of f and A. Expressed in terms of animal density, (i.e., $D = \mu_N/A$), the expected handling cost per plot per day is then proportional to DAp. If the capture probability is approximately proportional to trapping intensity, as when (2.13) applies and catchability (c) is low so that $p \approx cf/A$, then DAp may be replaced by the simpler expression Dcf. Multiplying the labor Dcf by the number of sites and days of trapping, costs are again proportional to ndf.

Intraplot travel distance per plot per day was seen (Figure 2.9) to be proportional to $\sqrt{Af}$, and hence total intraplot costs may be expressed as

$$C_4 nf + C_5 ndf + C_6 n\sqrt{Af} + C_7 nd\sqrt{Af}. \tag{2.35}$$

The number of treatments or treatment levels has been treated as a predetermined constant and is absorbed into these C-coefficients, and all treatments including the controls have been assumed to be equally replicated in a balanced design. Note that this intraplot contribution (2.35) to total cost results from four sources:

$$\begin{bmatrix} \text{Trap} \\ \text{costs} \end{bmatrix} + \begin{bmatrix} \text{Daily maintenance} \\ \text{of traps} \end{bmatrix} + \begin{bmatrix} \text{Intraplot travel during} \\ \text{preparation and cleanup} \end{bmatrix} + \begin{bmatrix} \text{Intraplot travel} \\ \text{during surveys} \end{bmatrix}.$$

The fine structure of expression (2.35) may change depending on the sampling and estimation strategy employed. For example, mark–recapture versus removal sampling results in somewhat different cost functions. Skalski (1985b) discusses these matters in detail; the present formulation of expression (2.35) represents costs associated with a mark–recapture study.

Form of Cost Function
The combined cost of fieldwork, including both fixed and variable costs, is now expressible as the sum of the two component models (2.34) and (2.35) adding to the net cost C_0, where

$$C_0 = C_1 n + C_2\sqrt{n} + C_3 d\sqrt{n} + C_4 nf + C_5 ndf + C_6 n\sqrt{Af} + C_7 nd\sqrt{Af}. \tag{2.36}$$

Simplifications of this general cost model for a mark–recapture study arise as special cases. For a single-plot abundance study, for example, setting $n = 1$ gives

$$C_0 = C_1 d + C_2 f + C_3 df + C_4\sqrt{Af} + C_5 d\sqrt{Af} \tag{2.37}$$

with appropriate adjustments to the C-coefficients. Further simplification results if the number d of trapping periods is predetermined and not available for manipulation. Absorbing this fixed value into the other constants gives the reduced formula

$$C_0 = C_1 f + C_2\sqrt{Af} \tag{2.38}$$

indicating economics of a single-plot study depend only on trapping effort and plot size.

EXAMPLE   The costs of a single-plot study will be used to illustrate the construction of a cost function (2.37). It will be assumed a budget of $5000 is available to establish and conduct a mark–recapture survey of small mammal abundance. The objective of constructing the cost function is identification of alternative levels of effort (i.e., d, f, and A) that are affordable within the $C_0 = \$5000$ budget. Site selection and characterization costs are anticipated to require 2 days of labor at $160 per day per person, resulting in an adjusted fixed cost of

$$C_0 = \$5000 - 2(160) = \$4680.$$

Costs that are strictly dependent on the number of days of trapping include such items as daily transportation costs and room and board for the field crew. It will be assumed that the site is located locally, precluding the need for personal living expenses. However, vehicle fees, including gasoline, are projected to be $60 per day. Therefore, allowing for 2 days of site selection and setup, 1 day of site cleanup, and d days of trapping, the interplot travel costs are described as $3(\$60) + \$60d$, where $C_1 = \$60$.

Cost on a per-trap basis are anticipated to include the purchase price per trap ($7.50), bait and bedding for warmth per trap ($0.25), plus setup costs. A two-person field crew is expected to be able to set up 400 traps during a day's labor. The labor costs for trap grid setup can therefore be expressed as

$$\frac{2(\$160)}{400} = \$0.80 \text{ per trap.}$$

The total trap costs on a per unit basis is then $\$8.55\,f$, where $C_2 = \$8.55$.

The time spent handling, marking, and recording data on the captured animals trapped per day, in turn, is a function of animal abundance (N) and trapping effort (f). Consequently, in constructing a cost function, animal density information is essential. It will be assumed that small-mammal density (D) is 50/ha and that the catch coefficient anticipated is $c = 0.005$. Using the approximate capture probability model if $p \approx cf/A$, the expected number of animals captured per day is then

$$DA\left(\frac{cf}{A}\right) = 50A\left(\frac{0.005f}{A}\right) = 0.25f.$$

For proper handling of the animals, it is necessary to tend the trap grid within the first 4 hours following sunrise. As such, the functional workday is 4 hours long, with a cost rate of $160/4 = \$40$/hour. This adjusted salary rate is an approximate cost of supplying a sufficient field crew to complete the work in 4 hours or less per day. Assuming handling time will be, on the average, 6 minutes per animal per day, the cost of daily trap maintenance is then

$$\left(\frac{\$160}{4}\right)d(0.25f)\left(\frac{6}{60}\right) = \$1\ df$$

where $C_3 = \$1.00$.

The costs of setting up the trap grid at the beginning of a d-day mark–recapture survey and collecting the equipment afterward includes intraplot travel costs. The rate of travel in setting up the trap grid is expected to be 1.25 km of trap line per hour. The costs of initially locating the traps before the first day of trapping is then

$$\left(\frac{\$160}{8}\right)\left(\frac{1}{1.25}\right)\sqrt{Af} = \$16\sqrt{Af}\,.$$

Travel during the removal of the trap grid and site cleanup is anticipated to be faster at a rate of 2.5 km of trap line per hour with a cost of

$$\left(\frac{\$160}{8}\right)\left(\frac{1}{2.5}\right)\sqrt{Af} = \$8\sqrt{Af}\,.$$

Hence, the intraplot travel costs during preparation and cleanup is therefore expressed as $\$24\sqrt{Af}$ where $C_4 = \$24$.

The intraplot travel during the d days of trapping can again be expressed by the rate the trap line is traversed. Recalling that the functional workday is reduced to 4 hours with a resulting labor rate of \$40/hour, and assuming a travel speed of 0.833 km of trap line per hour, the intraplot travel over the d days of trapping is expressed as

$$\left(\frac{\$160}{4}\right)\left(\frac{1}{0.833}\right)d\sqrt{Af} = \$48d\sqrt{Af}$$

where $C_5 = \$48.00$.

The single-plot cost function is then a result of the six coefficients calculated above where

$$\$4500 = \$60d + \$8.55f + \$1df + \$24\sqrt{Af} + \$48d\sqrt{Af}.$$

Fixing the trapping study to one work week (ie., $d = 5$), a cost function of the form (2.38) results, where

$$\$4200 = \$13.55f + \$264\sqrt{Af}.$$

For instance, with $f = 100$ traps, the affordable plot size is calculated to be a 1.2 ha with an anticipated small-mammal abundance of $N = 1.2(50) = 60$. Alternatively, with $f = 200$ traps, the affordable plot size is $A = 0.16$ ha with an anticipated small-mammal abundance of $N = 0.16(50) = 8$. Determination of the optimal combination of f and A requires consideration of sampling precision which will be a topic of Chapter 3.

Recommendations

Some of the factors such as capture probabilities and population size that influence the success of field investigations are familiar to investigators who have used mark–recapture methods. However, many more considerations are necessary in designing multiplot field studies to detect effects on mobile species. Some of these factors, such as the estimation of variance components and correlations, may be relatively new to many investigators. Therefore, preliminary surveys provide an important research tool in the exploration of population phenomenon and the design of experiments. Indeed, a history of preliminary survey data (as well as complete analyses of the consummate experiments) can provide insights into the spatial organization of populations not obtainable from any single investigations (e.g., Figure 2.6). Field biologists with a history of performing wildlife investigations should consider a research file on observed variance components and cost data as an important reference source and as an integral component in performing wildlife research.

Unless prior estimates of population parameters are available and reliable, we recommend that preliminary surveys be viewed as an initial step in most field investigations planning to use capture data to make inferences to population effect. Insights into the proper allocation of sampling efforts typically more than compensate for the reduction in an overall research budget resulting from the costs of the preliminary survey. The reason is that the remaining resources can be allocated in a more efficient manner to reduce error sources influencing the precision of important population contrasts.

Some recommendations for the design and analysis of preliminary surveys are as follows:

1. Hypotheses and appropriate test statistics should be specified prior to the preliminary survey to ensure that proper variance components and related parameters are estimated.
2. Cost data should be collected as an integral part of any preliminary survey.
3. Cost functions should be tentatively modeled before the preliminary survey to ensure that all necessary expenses are measured.
4. Available data sets should be reviewed and analyzed to provide prior values for parameters that cannot be readily obtained from a preliminary survey (i.e., temporal variance and covariances).
5. Researchers should maintain records of variance component analyses and cost data for designing future field trials.
6. Records on variance component analysis should be reviewed periodically to discern empirical and theoretical relationships about the spatial organization of wild populations.

We contend that these procedures have a place in most field experiments and are not limited to wildlife or mark–recapture investigatons. Rather, here we have attempted to illustrate how fundamental relationships in wildlife investigations can be discerned from preliminary data. These relationships will now be used in Chapters 3–6 to assist in design choices for wildlife investigations.

3

Surveys of Animal Abundance

In demographic studies, animal abundance is often only one parameter of interest in the study of population processes. An assessment of the viability of a wild population typically requires the characterization of the age structure of the population, as well as the estimation of natality rates and age-specific mortality rates. Mark–recapture methods such as those of the Jolly (1965)–Seber (1965) model (Seber 1982, pp. 196–205) or the general model of Robson (1969) can provide estimates of animal abundance, recruitment, and survival rates. Auxiliary information from the captured animals can then lead to additional estimates of the age and sex structure of the population. Recent work by Pollock (1982) and Kremers (1984) suggests, however, that there may be advantages in separating the task of the abundance estimation from the estimation of related parameters. Besides the potential model incompatibility of these objectives (Kremers 1984), Pollock (1982) suggests that a dual analysis using closed-population models for abundance estimation and release–recapture models for the estimation of survival rates may be more robust to the heterogeneity of individual capture probabilities.

A consequence of the multidimensional aspect of demographic studies is the need to assure sufficient precision for all important demographic parameters estimated. The traditional approach to the design of a multipurpose study is to design the study either (1) to have sufficient precision for the foremost parameter of interest or (2) to provide sufficient precision for the most variable parameter of interest. The effect of these approaches is to translate a basically multidimensional design problem to that of sample size determination in a univariate case. In the following discussion of population surveys, we will continue this tradition and assume that animal abundance is the parameter of primary interest. It should not be forgotten during the course of this discussion that population estimation may be only one of several objectives of a demographic study.

The estimation of animal abundance seeks to provide a measure of population size in the absence of complete enumeration. To accomplish this goal, mark–recapture data are used to model the capture probabilities of the animals, and in so doing, provide an estimate of animal abundance. Because an estimate of abundance is model-dependent, the most desirable estimation approach would simultaneously possess the properties of realism, specificity, and generality. However, no single estimator can be expected to be appropriate in all circumstances and for all animal populations. Consequently, if a survey of population abundance is to provide a valid estimate, capture data must conform to the analysis of various survey models.

Because the validity of a survey model can only be determined *a posteriori*, there is little alternative to the use of model selection to ensure validity of subsequent abundance estimates. The monographs by Otis et al. (1978) and Pollock et al. (1990) describe model selection in conjunction with multiple mark–recapture data. Similarly, Skalski and Robson (1982) provide a series of alternative estimators for abundance studies using sequential mark–recapture and removal trapping. Furthermore, Cormack (1985) presents the important model building advantages of generalized linear models (GLIM) in estimating animal abundance in both open and closed populations. We strongly recommend the analytic capabilities of these model selection procedures whenever the objective of a field study is abundance estimation. For demographic studies designed to have repeated surveys of abundance through time, independent model selection should be used at each of the separate sampling periods. Comparisons of abundance through time should be based on the most appropriate survey models for each survey period.

Survey Models

Because the focus of the book is on the design of wildlife investigations and not abundance estimation, three parsimonious survey models have been chosen to illustrate design concepts. *A priori*, there is no reasonable basis for calculating sample sizes using a complex survey model with a multiplicity of capture parameters. Usually, neither intuition nor past experience will be sufficient to predict distinct per period capture probabilities or quantify the nature of any heterogeneity in capture probabilities to be expected. Instead, the three survey models—(1) Lincoln Index, (2) constant probability, multiple-mark–recapture model [Darroch 1958, Otis et al. 1978 (i.e., model M_o)], and (3) constant probability, removal technique [Zippin 1956, 1958; Otis et al. 1978 (i.e., model M_B)]—have been selected for *a priori* design of surveys. Postdata analysis should then be based on model selection techniques.

Lincoln Index

The simple structure of the Lincoln Index and its potential application to both closed and open populations (when either natality or mortality but not both processes are occurring) suggests its broad appeal. Furthermore, the estimation of abundance for open populations using multiple recapture techniques in the presence of both natality and mortality developed by Manly and Parr (1968) is a special case of the Lincoln Index. Hence, the parsimonious and versatile Lincoln Index is a useful starting point for the design of population surveys.

The assumptions of the two-sample, mark–recapture method leading to the Lincoln Index are

1. There are equal and independent capture probabilities for all animals within a sampling period.
2. Marking does not affect catchability (implied by assumption 1 above).
3. Animals do not lose their marks.
4. All marked animals captured in the second period are reported.
5. Animals are randomly sampled either in both periods, or systematically after random mixing of marked and unmarked animals.
6. Population of size N is closed, or alternatively, if mortality only is occurring, $\hat{N}$ estimates population abundance at time of the first sample or if recruitment only is occurring, $\hat{N}$ estimates population abundance at time of second sample; if both mortality and recruitment are occurring, $\hat{N}$ is invalid.

Based on these assumptions and letting sample sizes during the two sampling periods (i.e., n_1 and n_2) be random variables, a multinomial likelihood model is implied.

The multinomial likelihood function for the capture data of an unconditional Lincoln Index is

$$L(n_1, n_2, m \mid p_1, p_2, N)$$

$$= \frac{N!}{m!(n_1 - m)!(n_2 - m)!(N - r)!}(p_1 p_2)^m (p_1 q_2)^{n_1 - m}(q_1 p_2)^{n_2 - m}(q_1 q_2)^{N - r}$$

$$(3.1)$$

where N = animal abundance

$p_1 = (1 - q_1)$ = probability of capture during initial sample

$p_2 = (1 - q_2)$ = probability of capture during recapture period

n_1 = number of animals captured and released during initial sample

n_2 = number of animals captured in the second sampling period

m = number of animals marked and recaptured

$r = n_1 + n_2 - m$ = number of distinct animals captured from the population

The maximum likelihood estimators for the parameters in model (3.1) are

$$\hat{N} = \frac{n_1 n_2}{m} = \frac{r}{1 - (1 - \hat{p}_1)(1 - \hat{p}_2)} \tag{3.2}$$

$$\hat{p}_1 = \frac{m}{n_2} \tag{3.3}$$

$$\hat{p}_2 = \frac{m}{n_1}. \tag{3.4}$$

Chapman (1951) provides an adjustment for bias with the formula

$$\hat{N} = \frac{(n_1 + 1)(n_2 + 1)}{(m + 1)} - 1. \tag{3.5}$$

The adjusted Lincoln Index is well defined (finite) for all outcomes (n_1, n_2, m) and is conditionally unbiased when $n_1 + n_2 > N$ but unconditionally is negatively biased by the factor

$$\% \text{ bias} = -q_1 q_2 (1 - p_1 p_2)^{N-1} \, 100\%$$

for fixed N. When animal abundance (N) is negative binomially distributed with parameters $P(= 1 - Q)$ and R, the relative bias has the expected value

$$E_N[\% \text{ bias}] = \frac{-q_1 q_2}{(1 - p_1 p_2)} \left(\frac{P}{1 - Q(1 - p_1 p_2)} \right)^R \, 100\%.$$

Similarly, the expected bias is

$$E_N[\% \text{ bias}] = \frac{-q_1 q_2}{(1 - p_1 p_2)} e^{-\mu p_1 p_2} 100\%$$

when N is Poisson distributed with mean μ.

The variance of the adjusted Lincoln Index (3.5) can be expressed as

$$\text{Var}(\hat{N} \mid N) = N^2 \left(\frac{N}{n_1 n_2} + 2 \left(\frac{N}{n_1 n_2} \right)^2 + 6 \left(\frac{N}{n_1 n_2} \right)^3 \right) \tag{3.6}$$

with variance estimate

$$\hat{\text{Var}}(\hat{N} \mid N) = \frac{(n_1 + 1)(n_2 + 1)(n_1 - m)(n_2 - m)}{(m + 1)^2(m + 2)}. \tag{3.7}$$

The variance of the maximum likelihood estimate (3.2), expressed as a function of capture probabilities and useful for subsequent design of surveys, is

$$\text{Var}(\hat{N} \mid N) = \frac{Nq_1q_2}{p_1p_2} \tag{3.8}$$

where $q_i = 1 - p_i$ for $i = 1, 2$. The conditional variance of the catch index (r), expressed as a function of the capture probabilities, is

$$\text{Var}(r \mid N) = Nq_1q_2(1 - q_1q_2) \tag{3.9}$$

with a conditional expected value

$$E(r \mid N) = N(1 - q_1q_2). \tag{3.10}$$

In Chapters 4–6, where the purpose of the wildlife investigations is the comparison of animal abundance, the relative magnitudes of the sampling errors of $\hat{N}$ and r will be of primary importance. The potential use of catch index (r) in population comparisons is motivated by the inherently smaller sampling error associated with the measure of relative abundance (r) than absolute abundance ($\hat{N}$). To see this relationship, a comparison of the coefficients of variation [i.e., $CV = \sqrt{\text{Var}(\hat{\theta})}/E(\hat{\theta})$] for r and $\hat{N}$ show that

$$CV(r \mid N) < CV(\hat{N} \mid N)$$

$$\frac{\sqrt{N(1 - q_1q_2)q_1q_2}}{N(1 - q_1q_2)} < \frac{\sqrt{Nq_1q_2/(p_1p_2)}}{N}$$

$$\frac{\sqrt{1 - q_1q_2}}{(1 - q_1q_2)} < \sqrt{\frac{1}{p_1p_2}}$$

$$p_1p_2 < 1 - q_1q_2$$

$$2 < \frac{1}{p_1} + \frac{1}{p_2}$$

which holds whenever the capture probabilities (i.e., p_1 and p_2) are <1. A strict equality exists [i.e., $CV(r \mid N) = CV(\hat{N} \mid N)$] in the sampling precision of the measures of relative abundance and absolute abundance only when populations are completely enumerated (i.e., $p_1 = p_2 = 1$). Similar arguments can be set forth for the other survey models presented below.

The catch index r is a sufficient summary for abundance in tests of hypotheses if capture probabilities are homogeneous. The choice of whether to use $\hat{N}$ or r in population comparisons will be dependent on whether capture probabilities are homogeneous between populations being compared. A test of homogeneous capture probabilities $[\underline{p} = (p_1, p_2)]$ among populations surveyed by Lincoln Indices can be constructed from a multinomial likelihood function conditional on the number of distinct animals captured, r, for each population. The quantities m, $n_1 - m$, $n_2 - m$ are conditionally trinomially distributed and can be represented by a $3 \times C$ contingency table as follows.

Capture history	Population 1	Population 2	. . .	Population C	
Caught in both samples	m_1	m_2	. . .	m_C	$m.$*
Caught only in first sample	$n_{11} - m_1$	$n_{21} - m_2$	. . .	$n_{C1} - m_C$	$n._1 - m.$
Caught only in second sample	$n_{12} - m_1$	$n_{22} - m_2$	. . .	$n_{C2} - m_C$	$n._2 - m.$
	r_1	r_2	. . .	r_C	$r.$

(3.11)

*Dots denote summation across populations.

A chi-square test of homogeneity (Snedecor and Cochran 1980, pp. 208–210) for the $3 \times C$ table with $2(C - 1)$ degrees of freedom can serve as a test of the hypothesis

$$H_0: (p_{i1}, p_{i2}) = (p_1, p_2) \quad \text{for} \quad i = 1, \ldots, C$$

against

$$H_a: (p_{i1}, p_{i2}) \neq (p_1, p_2) \quad \text{for some} \quad i = 1, \ldots, C.$$

It is important to note that contingency table (3.11) remains valid despite the possibility that capture probabilities $[(p_{1j}, p_{2j})$ for $j = 1, \ldots, N]$ among the N individual animals of a population are not homogeneous (Table 3.1). The reason contingency table (3.11) remains valid pertains to the nature of the hypothesis tested. The chi-square statistic does not test whether the frequency of individual capture potentials are identical among the populations being compared. Rather, contingency table (3.11) tests whether the capture probabilities of the individual animals within a population $[(p_{1j}, p_{2j})]$, as observed through capture histories, are a sample from a single common distribution or samples from different distributions for each population. Viewed from this perspective of an infinite process, all N animals of a population then have the same expected capture potential prior to their realization. As such, contingency table (3.11) which is based on model M_T (Otis et al. 1978) remains valid even though realized capture potentials may differ among animals. Consequently, contingency table (3.11) can be

Table 3.1

Observed α-Levels for χ^2 Tests of a 3 $\times$ 2 Contingency Table for Interpopulation
Homogeneity (3.11) Where Individual Capture Probabilities Are Beta (s, λ)-Distributed
in Accordance with Model M_H^*

Abundance		Beta		α-Level		
μ_N	σ_N^2	s	λ	0.10	0.05	0.01
50	100	0.5	0.5	0.1095	0.0555	0.0110
		1.0	1.0	0.0970	0.0495	0.0095
		3.0	3.0	0.1025	0.0530	0.0095
		0.6	1.4	0.0990	0.0470	0.0080
		3.0	7.0	0.1050	0.0520	0.0095
		0.42	0.98	0.1110	0.0520	0.0095
200	1000	0.5	0.5	0.1130	0.0570	0.0085
		1.0	1.0	0.0960	0.0460	0.0060
		3.0	3.0	0.0970	0.0490	0.0085
		0.6	1.4	0.1015	0.0495	0.0105
		3.0	7.0	0.0970	0.0475	0.0105
		0.42	0.98	0.1015	0.0465	0.0050

*Otis et al. 1978. Animal abundance was assumed to be negative-binomially distributed
with μ_N and σ_N^2 for both populations tested. Results are based on 2000 computer simulations
for each scenario.

extended for use with multiple-mark–recapture surveys to detect interpopulation heterogeneity under all eight model scenarios discussed in Otis et al.
(1978).

Before proceeding, a final comment concerning contingency table
(3.11) is necessary. The role of tests of homogeneity in field experiments
can be viewed as checking the calibration of the replicate mark–recapture
surveys. From the simple model (3.1) where

$$E(r \mid N) = N(1 - q_1 q_2)$$

and

$$E(\hat{N} \mid N) \approx N(1 - e^{-Np_1p_2})$$

it becomes apparent that the tests of homogeneity do not require strict proportionality to perform this implicit calibration. In the presence of intrapopulation heterogeneity, the relationship between r (or $\hat{N}$) and N and p becomes even more complex, yet contingency table (3.11) remains valid. This
is unlike ratio and regression estimators (Cochran 1977, pp. 150–153,
189–190), which provide a means of explicit calibration based on an assumption of linearity. Consequently, the tests of homogeneity can be
viewed as determining whether the nonlinear functions of r or $\hat{N}$ in terms of
N and p are the same among the populations being compared. As such, the
capture data from mark–recapture surveys provide a more realistic and

complex check of calibration than possible using standard calibration techniques.

Darroch's Constant Probability Multiple-Mark–Recapture Model

The most parsimonious multiple-mark–recapture model is the constant probability model developed by Darroch (1958) and later denoted as model M_o by Otis et al. (1978). The survey technique consists of k survey periods. In the initial sampling period, n_1 animals are captured, marked, and released back into the population. Ideally, for the capability of subsequent model selection, animals would be marked with uniquely numbered tags. In subsequent sampling periods, the $n_i (i = 2, \ldots, k)$ captured animals would be recorded as either previously marked or as unmarked. Unmarked animals would be marked and all n_i animals released into the population.

The key assumptions of the Darroch (1958) model are

1. The population of size N is closed.
2. All N animals have independent and equal probabilities of capture in the ith period $(i = 1, \ldots, k)$.
3. Marking has no effect on an animal's subsequent probability of capture (implied by point 2 above).
4. Animals do not lose their marks.
5. The probability of capture is constant during the k sampling periods.

These five assumptions lead to the multinomial likelihood function

$$L(f_1, \ldots, f_k \mid N, p)$$

$$\propto \binom{N}{f_1, f_2, \ldots, f_k, N - r}(p(1 - p)^{k-1})^{f_1}(p^2(1 - p)^{k-2})^{f_2} \ldots (p^k)^{f_k}((1 - p)^k)^{N-r}$$

$$(3.12)$$

where f_i = number of animals captured i times $(i = 1, \ldots, k)$
$\quad\quad r = \Sigma_{i=1}^{k} f_i$ = number of distinct animals captured during the survey
$\quad\quad p$ = per-period probability of capture

The likelihood can be further simplified by defining the quantity $n. = \Sigma_{i=1}^{k} n_i = \Sigma_{i=1}^{k} if_i$ = total number of captures during the survey, in which case

$$L(f_1, \ldots, f_k \mid N, p) \propto \binom{N}{f_1, f_2, \ldots, f_k, N-r} p^{n.}(1 - p)^{kN-n.}.$$

The maximum likelihood estimator for N based on the Darroch model is the unique solution to the equation

$$\left(1 - \frac{r}{N}\right) = \left(1 - \frac{n.}{kN}\right)^k. \tag{3.13}$$

Darroch (1958) gives the asymptotic variance of the estimator $\hat{N}$ of

$$\text{Var}(\hat{N} \mid N) = N[(1 - p)^{-k} - k(1 - p)^{-1} + k - 1]^{-1}. \tag{3.14}$$

The variance can be estimated by substituting $\hat{N}$ and $\hat{p}$ into (3.14) where

$$\hat{p} = \frac{n.}{k\hat{N}}.$$

As in the case of the Lincoln Index, the catch index (r) provides an alternative metric for population comparison (Chapters 4–6) should capture probabilities be homogeneous between the populations being compared. The expected value of the catch index for the Darroch model conditional on population abundance is

$$E(r \mid N) = N(1 - (1 - p)^k)$$

with associated conditional variance

$$\text{Var}(r \mid N) = N(1 - (1 - p)^k)(1 - p)^k.$$

A test of homogeneous capture probabilities between populations surveyed under the Darroch model can be constructed using the multinomial likelihood (3.12) conditional on the numbers of distinct animals captured per population (i.e., r values). The test of homogeneity can be represented by a $k \times C$ contingency table of the form

Capture frequency	Population 1	Population 2	. . .	Population C	
Caught once	f_{11}	f_{21}	. . .	f_{c1}	$f_{.1}$
Caught twice	f_{12}	f_{22}	. . .	f_{c2}	$f_{.2}$
$\vdots$	$\vdots$	$\vdots$	$\vdots$	$\vdots$	$\vdots$
Caught k times	f_{1k}	f_{2k}	. . .	f_{ck}	$f_{.k}$
	r_1	r_2	. . .	r_c	$r.$

$$\tag{3.15}$$

The chi-square test of homogeneity with $(k - 1)(C - 1)$ degrees of freedom can be used to test the assumption of interpopulation homogeneity in the context of multiple-mark–recapture surveys. The test of homogeneity will remain valid despite the nature of trap–response within the populations. Consequently, the test statistic can be used under all eight models for heterogeneity in trap response presented in Otis et al. (1978) for multiple-recapture models. The contingency table analysis (3.15) is an alternative to a competing contingency table analysis based on the $2^k - 1$ observable capture histories, of which contingency table (3.11) is a special case (i.e., for $k = 2$). When the capture data conforms to likelihood (3.12), the proposed $k \times C$ contingency table analysis (3.15) will have greater power. The table may also be used when a contingency table based on the $2^k - 1$ capture histories requires pooling because cell counts are too small. The best strategy is to use a contingency table that conforms most closely to the multinomial likelihood appropriate for the survey data.

Zippin's Constant Probability Removal Model

The third parsimonious survey model we will consider in the design of wildlife investigations is the constant probability removal technique of Zippin (1956, 1958). The survey consists of k consecutive sampling periods. Animals captured $(n_i, i = 1, \ldots, k)$ in each sampling period are removed from the population prior to the next sampling period. With the use of equal sampling effort during each period, the probability of captures may be homogeneous between periods resulting in a geometric sequence of counts over k sampling periods.

The assumptions of Zippin's (1956, 1958) constant probability removal technique are

1. The population of size N is closed.
2. All N animals have independent and equal probabilities of capture in the ith sampling period $(i = 1, \ldots, k)$.
3. Probabilities of capture are constant (i.e., $p_i = p$ for $i = 1, \ldots, k$) over the k sampling periods.

These three assumptions lead to the multinomial likelihood function

$$L(n_1, \ldots, n_k \mid N, p) = \binom{N}{n_1, \ldots, n_k, N - r} p^{n_1}(p(1 - p))^{n_2} \ldots$$
$$(p(1 - p)^{k-1})^{n_k}(1 - p)^{(N-r)k} \qquad (3.16)$$

where $r = \sum_{i=1}^{k} n_i$. Defining $t_2 = \sum_{i=2}^{k} (i - 1)n_i$, we can simplify the likelihood to

$$L(\underline{n}_i \mid N, p) = \binom{N}{n_1, \ldots, n_k, (N - r)} p^r(1 - p)^{k(N-r)+t_2}.$$

The quantities r and t_2 are minimum sufficient statistics for the likelihood (3.16).

The maximum likelihood estimators (MLEs) for N and p are

$$\hat{N} = \frac{r}{1 - (1 - \hat{p})^k} \qquad (3.17)$$

where $\hat{p}$ is the solution to the equation

$$\frac{1 - \hat{p}}{\hat{p}} - \frac{k(1 - \hat{p})^k}{1 - (1 - \hat{p})^k} = \frac{t_2}{r}.$$

The asymptotic variances for the MLEs are

$$\text{Var}(\hat{N} \mid N) = \frac{N(1 - q^k)q^k}{(1 - q^k)^2 - (pk)^2 q^{k-1}} \qquad (3.18)$$

and

$$\text{Var}(\hat{p} \mid p) = \frac{(qp)^2(1 - q^k)}{N[q(1 - q^k)^2 - (pk)^2 q^k]} \qquad (3.19)$$

where $q = 1 - p$. Variance estimates are calculated by substituting $\hat{N}$ and $\hat{p}$ for the parameters in Eqs. (3.18) and (3.19).

The statistic r for the Zippin model again provides a catch index for relative abundance. The conditional expected value and variance of r under the constant probability removal technique are, respectively

$$E(r \mid N) = N(1 - (1 - p)^k)$$

and

$$\text{Var}(r \mid N) = N(1 - (1 - p)^k)(1 - p)^k.$$

A test of homogeneous capture probabilities between populations surveyed under the Zippin model can be based on a contingency table analysis and the multinomial categories found in likelihood (3.16). The test of homogeneity can be represented by a $k \times C$ contingency table of the form

Caught in period	Population 1	Population 2	. . .	Population C	
1	n_{11}	n_{21}	. . .	n_{c1}	$n_{\cdot 1}$
2	n_{12}	n_{22}	. . .	n_{c2}	$n_{\cdot 2}$
$\vdots$	$\vdots$	$\vdots$	. . .	$\vdots$	$\vdots$
k	n_{1k}	n_{2k}	. . .	n_{ck}	$n_{\cdot k}$
	r_1	r_2	. . .	r_c	$r.$

(3.20)

The chi-square statistic for the test of homogeneity of table (3.20) has $(k - 1)(C - 1)$ degrees of freedom. The contingency table analysis remains valid despite violations of the assumptions of independent, identical, and constant probabilities of capture. When the Zippin model is valid [see Skalski and Robson (1979) for goodness-of-fit test], a more powerful test of homogeneity can be performed comparing the t_2 values to their expected values conditional on the row and column totals of contingency table

(3.20). In the case of two populations, the statistic for the test of homogeneity is

$$Q = \frac{\left(\dfrac{t_{21}}{r_1} - \dfrac{t_{22}}{r_2}\right)^2}{S^2\left(\dfrac{1}{r_1} + \dfrac{1}{r_2}\right)} \tag{3.21}$$

where

$$S^2 = \frac{\sum\limits_{i=2}^{k} (i-1)^2 n_{\cdot i} - \dfrac{(t_{2\cdot})^2}{r_{\cdot}}}{r_{\cdot} - 1}$$

and where

$$t_{2\cdot} = \sum_{j=1}^{2} t_{2j}$$

$$r_{\cdot} = \sum_{j=1}^{2} r_j \quad .$$

Under the null hypothesis of interpopulation homogeneity, the Q-statistic is chi-square-distributed with 1 degree of freedom (Skalski and Robson 1979). The power to detect interpopulation heterogeneity can be calculated by the noncentrality parameter presented in Skalski et al. (1984).

For multiple populations, the Q-statistic can be extended (Skalski and Robson 1979) to test for homogeneity between C populations. The chi-square test for C populations can be written as

$$Q = \frac{t_{2\cdot}}{H} \sum_{j=1}^{C} \frac{(t_{2j} - f_j t_{2\cdot})^2}{f_j t_{2\cdot}} \tag{3.22}$$

with

$$t_{2\cdot} = \sum_{j=1}^{C} t_{2j},$$

$$f_j = \frac{r_j}{T},$$

$$T = \sum_{j=1}^{C} r_j,$$

and

$$H = \frac{T\left(\sum\limits_{i=2}^{k} (i-1)^2 n_{\cdot i}\right) - \left(\sum\limits_{i=2}^{k} (i-1) n_{\cdot i}\right)^2}{(T-1)} .$$

The test statistic is approximately distributed as a chi-square with $(C - 1)$ degrees of freedom.

The versatility of the Zippin model for the estimation of animal abundance extends to its use in the analysis of multiple-recapture data. Otis et al. (1978) apply the use of the Zippin estimator to multiple-recapture models M_b and M_{bh}, where capture probabilities change as a consequence of prior capture (i.e., M_b) and where capture probabilities vary between animals and in the presence of trap response (i.e., M_{bh}). The latter application of the Zippin estimator to model M_{bh} has been termed the "generalized removal technique." Tests of goodness of fit to the generalized removal technique are presented in Otis et al. (1978) with tests specific to the likelihood model (3.16) presented in Skalski and Robson (1979).

Distributional Properties of Abundance Estimators

The estimation of animal abundance $(\hat{N})$ and its associated variance estimate $[\text{Vâr}(\hat{N} \mid N)]$ based on a viable survey model is a necessary first step in the process of statistical inference. However, the variance estimate $[\text{Vâr}(\hat{N} \mid N)]$ may not be very descriptive or useful in constructing probabilistic statements concerning animal numbers (Sprott 1981). In skewed sampling distributions, the variance estimate gives little or no information on the asymmetry of the distribution. Without this knowledge, confidence interval estimates may be in error much more frequently than realized.

The theory of maximum likelihood estimation [MLE; for a nontechnical discussion as it pertains to abundance estimation, see Otis et al. (1978, pp. 102–105)] ascribes to population estimators a number of desirable properties: (1) as sample size (r) approaches N, the estimator converges to the true population size [i.e., consistency, $P(\lim_{r \to N} \hat{N} = N) = 1$]; (2) asymptotic normality; (3) asymptotic sufficiency; and (4) an asymptotic variance equal to the minimum variance bound. All these properties are attained as sample sizes and population levels increase without bound. Unfortunately, there is no good way to know when animal abundance (N) or sample sizes are sufficiently large for these properties of an abundance estimator to be practically assumed.

All else being equal, abundance estimates from a population of size $N = 500$ will be more normally distributed than estimates from a population of 50 animals. Whether a population size of 500 is sufficiently large to viably assume approximate normality, however, remains unknown. A normally distributed random variable has a continuous range of values from minus to plus infinity $(-\infty$ to $+\infty)$. Population estimates, on the other hand, have a restricted range between 0 (more often $\geq r$, the number of individuals captured in a closed population) and $+\infty$, consequently, they never can be precisely normally distributed. The effect is that confidence interval construction of the form

$$\hat{N} \pm Z_{1-\alpha/2}\sqrt{\hat{\text{Var}}(\hat{N} \mid N)} \qquad (3.23)$$

will be at best only approximately correct, or more exactly, asymptotically correct.

Although the interval estimate (3.23) performs reasonably well under most circumstances, use of data transformations often can result in a variate which is more normally distributed. The principal way this occurs is by a more symmetric distribution for the transformed $\hat{N}$ (Figure 3.1). Consequently, the resulting interval estimates based on the assumption of normality will more closely approximate the nominal $(1 - \alpha)$ levels for both one-tailed and two-tailed confidence limits under transformation.

Various transformations have been suggested for use with abundance estimates. Cormack (1968, p. 461) and Ricker (1958) have suggested that $1/\hat{N}$ has a more symmetric distribution than $\hat{N}$ for a Lincoln Index. Alternatively, Skalski (1985a) found from simulation studies that $\ln \hat{N}$ is more normally distributed and will result in interval estimates with α-levels closer to nominal levels. For abundance estimates from removal sampling (Zippin 1956, 1958) or the generalized removal model (Otis et al. 1978, Skalski and Robson 1982), Skalski et al. (1984) found $1/\hat{N}$ to be more normally dis-

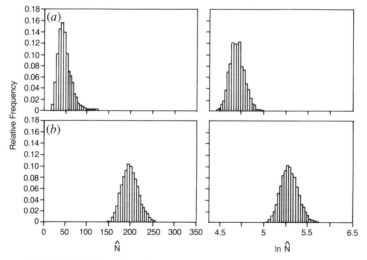

Figure 3.1 Distribution of the Chapman (1951) estimator for the Lincoln Index and its ln-transformation when the sampling variances are the same, $\text{Var}(\hat{N}) = 400$, but the population parameters differ, where (a) $N = 50$, $p_1 = p_2 = 0.2612$ and (b) $N = 200$, $p_1 = p_2 = 0.4142$.

tributed and to provide better interval estimates than several alternative transformations. In yet another case, Sprott (1981) suggested that the variate $\hat{N}^{-3}$ for the multiple-recapture method of Darroch and Ratcliff (1980) was approximately normally distributed and should be used in interval estimation with that model. In single-plot abundance studies, any transformation that results in near-normality may be selected for interval estimation and hypothesis testing. In the multiplot studies to be discussed later, the additional considerations of additivity and homoscedasticity must be included in the selection of a data transformation.

Statistical Analysis after Abundance Has Been Estimated

As the previous section has suggested, the analysis of capture data is not necessarily complete simply because animal abundance has been estimated and its variance computed. Inference from the capture data to the population sampled is necessary. Two alternative approaches can be taken at this point in the analysis: either an interval estimate of abundance may be constructed or a statistical test concerning abundance relationships may be performed. Historically, both approaches to the analysis of abundance studies have been used and sometimes abused.

Confidence Interval Estimation

Usually, the most appropriate analysis in conjunction with an abundance estimate is to construct as realistic confidence interval for population size. Let $T(\hat{N})$ represent an appropriate transformation of an abundance estimate which results in approximate normality of the resulting variate (e.g., ln-transformation of Lincoln Index.) Using the delta method (Seber 1982, pp. 7–9), the pivotal quantity

$$\frac{T(\hat{N}) - T(N)}{\sqrt{\text{V\^ar}(\hat{N})\left(\frac{dT(N)}{dN}\right)^2_{|\hat{N}}}} \tag{3.24}$$

is approximately distributed as a standard normal variable [$N(0, 1)$]. With the use of pivotal quantity (3.24), a two-tailed confidence interval for N can be constructed from the expression

$$P\left(Z_{\alpha/2} \leq \frac{T(\hat{N}) - T(N)}{\sqrt{\text{V\^ar}(\hat{N})\left(\frac{dT(N)}{dN}\right)^2_{|\hat{N}}}} \leq Z_{1-\alpha/2}\right) = 1 - \alpha \tag{3.25}$$

by solving for N. In the case of a 95% confidence interval, the values of $Z_{\alpha/2}$ and $Z_{1-\alpha/2}$ are the familiar quantities -1.96 and 1.96, respectively.

To express (3.25) in terms of N, the pivotal quantity within the brackets must be "inverted," in other words, solved as a function of N. For example, assuming a ln-transformation is appropriate, (3.25) becomes

$$P\left(Z_{\alpha/2} \leq \frac{\ln \hat{N} - \ln N}{\sqrt{\dfrac{\text{Vâr}(\hat{N})}{\hat{N}^2}}} \leq Z_{1-\alpha/2}\right) = 1 - \alpha$$

which can, in turn, be written as

$$P\left(\hat{N}e^{-Z_{1-\alpha/2}\sqrt{\text{Vâr}(\hat{N})/\hat{N}^2}} \leq N \leq \hat{N}e^{-Z_{\alpha/2}\sqrt{\text{Vâr}(\hat{N})/\hat{N}^2}}\right) = 1 - \alpha. \tag{3.26}$$

In the management and investigation of wild populations (e.g., setting harvest regulations, study of nutrient cycling), one-tailed confidence limits often may be of more practical value than the more generally used two-tailed interval estimates. Game managers, for instance, may be interested in estimating the minimum number of animals present on a refuge. Alternatively, a researcher may be interested in the maximum nutrient transport and consequently, an upper bound on population size. Using the pivotal quantity (3.24), a lower one-tailed interval estimate of N [i.e., $P(N \geq N_L) = 1 - \alpha$] can be found from the expression

$$P\left(Z_{1-\alpha} \geq \frac{T(\hat{N}) - T(N)}{\sqrt{\text{Vâr}(\hat{N})\left(\dfrac{dT(N)}{dN}\right)^2_{|\hat{N}}}}\right) = 1 - \alpha. \tag{3.27}$$

Similarly, an upper one-tailed interval estimate of N [i.e., $P(N \leq N_U) = 1 - \alpha$] can be calculated from the expression

$$P\left(Z_{\alpha} \leq \frac{T(\hat{N}) - T(N)}{\sqrt{\text{Vâr}(\hat{N})\left(\dfrac{dT(N)}{dN}\right)^2_{|\hat{N}}}}\right) = 1 - \alpha. \tag{3.28}$$

It is here in these calculations of one-tailed confidence intervals that the symmetry of the distribution for the transformed variate for $\hat{N}$ is particularly important. Although the actual α-levels for two-tailed interval estimates are usually very close to nominal levels with normal approximations such as (3.23) and (3.25), one-tailed intervals tend not be as well behaved. In extreme cases, all or most of the errors of exclusion from two-tailed confidence intervals can occur in a single tail. Consequently, a conservative interpretation of a $(1 - \alpha)$ one-tailed interval would be to admit the possibility of an error rate as high as 2α.

EXAMPLE Capture data from a single mark–recapture study of the Great Basin pocket mouse (*Perognathus parvus*) on a 1-ha study plot will be used to illustrate these computations. The study yielded $n_1 = 50$ marked animals during 5 days of trapping with $m = 38$ recaptures out of $n_2 = 50$ animals caught during a second 4-day capture period. The resulting point estimate of abundance based on a Lincoln Index (Chapman 1951) was $\hat{N} = 67.447$ ($r = 63$ distinct animals trapped) with an associated variance estimate of $\hat{Var}(\hat{N}\,|\,N) = 7.8054$. A 95% confidence interval estimate of the pocket mouse abundance at the study area can then be found from the expression

$$CI\left(-1.96 \leq \frac{4.2113 - \ln N}{\sqrt{\dfrac{7.8054}{(67.477)^2}}} \leq 1.96\right) = 0.95$$

or $CI(62.2 \leq N \leq 73.2) = 0.95$. The resultant interval estimate is asymmetric about the point estimate $\hat{N} = 67.4$ in lieu of a symmetric error rate (i.e., $\alpha/2$) with regard to over- and underestimating the true abundance level (N). As stated earlier, in closed-population studies, the lower tail can be truncated at the value r if (3.25) falls below that value. This behavior of interval estimates yielding inadmissible values of N is an artifact of the normal approximation used in interval construction, and not a fault of point estimation or study design as Otis et al. (1978) would suggest. The switch in notation to CI, an abbreviation for confidence interval, is used once numerical values are substituted into (3.25). The notational change is necessary, because following numerical evaluation, the interval estimate has either correctly or incorrectly included N.

To illustrate one-tailed interval estimation, the pocket mouse data yields an upper 95% one-tailed interval estimate for abundance of

$$CI\left(-1.645 \leq \frac{4.2113 - \ln N}{\sqrt{\dfrac{7.8054}{(67.477)^2}}}\right) = 0.95$$

or

$$CI(N \leq 72.2) = 0.95.$$

Because the earlier two-tailed confidence interval included inadmissible values of N (i.e., lower tail $N_L = 62.2 < 63$), the normal approximation should be viewed with suspicion and the one-tailed interval used cautiously.

Hypothesis Testing

The use of hypothesis testing in conjunction with a single abundance estimate has limited applicability. A null hypothesis of the form

$$H_0: N = N_0 \tag{3.29}$$

against

$$H_a: N \neq N_0$$

where N_0 is some specified numerical value has limited utility unless there is some special extrasurvey information that would lead one to believe that there should be **exactly** N_0 individuals in the population. In virtually all other circumstances, a simple hypothesis of the form $H_0: N = N_0$ is nonsensical.

Alternatively, a test of hypothesis may be a useful decision rule in making selections among qualitative statements (composite hypotheses) concerning animal abundance. For instance, management practices may state that hunting will be permitted only if local animal abundance exceeds some value N_0 determined by the carrying capacity of the range. In still other circumstances, management intervention may be indicated if local population size falls below or exceeds some range in values. One example might be at deer yards where either increased hunting pressure or improvements in winter range would be signaled by large departures from long-term trends in abundance levels. Such scenarios suggest three possible sets of alternative hypotheses of general applicability as follows:

$$H_0: N \geq N_0 \tag{3.30}$$
$$H_a: N < N_0,$$

or

$$H_0: N \leq N_0 \tag{3.31}$$
$$H_a: N > N_0,$$

or still

$$H_0: N_L \leq N \leq N_U \tag{3.32}$$
$$H_a: N < N_L \quad \text{or} \quad N > N_U.$$

The pivotal quantity (3.24) provides the basis for the test of these composite hypotheses, (3.30)–(3.32). Because there is a one-to-one correspondence between confidence interval construction and the appropriate test statistic, a test of hypotheses can be conducted directly from the interval estimates.

The one-tailed hypotheses (3.30) and (3.31) can be tested by observing whether the value of N_0 falls within the limits of the one-tailed interval estimates (3.28) and (3.27), respectively. If the value of N_0 does not lie within the observed confidence limits, the null hypothesis is rejected at a significance level α. Similarly, an α-level two-tailed test of the hypothesis (3.32) can be conducted using a two-tailed $(1 - \alpha)$ confidence interval from Eq. (3.25). The null hypothesis (3.32) is rejected in favor of its alternative if at least one of the values N_L or N_U falls outside the confidence limits.

The relationship between the calculation of confidence intervals and tests of hypotheses is evident from the discussion above. What may not be so evident are the inherent differences in the logic of the two approaches of analysis of the abundance data. In interval estimation, the confidence interval provides an expression of the relative certainty associated with our estimate of population size. The calculations are of an *a posteriori* nature for they are not dependent on prior conceptions concerning population size. On the other hand, tests of hypotheses are of an *a priori* nature with values of N_0, N_L, and N_U specified before the study is conducted. The choice of null hypothesis, (3.30), (3.31), or (3.32), and the specific endpoints chosen should result from the stated objectives of the investigation and not from the results of the field sampling. Formulation of hypotheses after animal abundance has been estimated, or confidence intervals constructed, changes the α-level unpredictably and violates the spirit of an objective evaluation of test results.

Specifying the Precision of Abundance Studies

The design of a mark–recapture study is as important to the eventual success of a population survey as is the subsequent data analysis. All too often, however, investigators take a fatalistic view of the design of abundance studies, believing that the results are largely outside their control. Although the success of a mark–recapture survey cannot be guaranteed, the anticipated performance of such a study can be predicted from knowledge of the anticipated population size and rates of capture. These predictions then can be compared with the desired performance of the study and the field design adjusted accordingly.

A useful measure of the success of an abundance study is the precision of the population estimate. Robson and Regier (1964) have proposed guidelines for the precision of abundance studies, but there are no hard-and-fast rules that are appropriate for all occasions. The only certainty is the need to consider the anticipated performance of the study before it is implemented. Otherwise, the fatalistic beliefs concerning the success of abundance studies are indeed self-fulfilling prophecies.

The coefficient of variation (CV) of an abundance estimator is a particularly useful and convenient measure of precision because it enables one to calculate the approximate probability of the prediction erring from N by any given percentage. If $CV(\hat{N})$ is 25%, for example, one can be approximately 95% sure that the estimator $\hat{N}$ does not err by more than 50%, and approximately 80% sure that $\hat{N}$ does not deviate from the true N by more than one-third of N. In general, the probability of not erring by more than $100\epsilon\%$,

$$P\left(\left|\frac{\hat{N} - N}{N}\right| < \epsilon\right) \geq 1 - \alpha, \tag{3.33}$$

is approximately

$$P(|\hat{N} - N| < N\epsilon) = P\left(\frac{|\hat{N} - N|}{\sqrt{\mathrm{Var}(\hat{N})}} < \frac{N\epsilon}{\sqrt{\mathrm{Var}(\hat{N})}}\right)$$
$$= P\left(\frac{|\hat{N} - N|}{\sqrt{\mathrm{Var}(\hat{N})}} < \frac{\epsilon}{CV(\hat{N})}\right)$$
$$\approx P\left(|Z| < \frac{\epsilon}{CV(\hat{N})}\right)$$

where Z denotes a standard normal deviate. Letting $\phi(Z)$ denote the cumulative standard normal probability distribution, we obtain

$$P\left(\left|\frac{\hat{N} - N}{N}\right| < \epsilon\right) \approx 1 - 2\phi\left(\frac{-\epsilon}{CV(\hat{N})}\right) \qquad (3.34)$$

where $CV(\hat{N})$ is measured in decimal units, or

$$P\left(\left|\frac{\hat{N} - N}{N}\right| < \epsilon\right) \approx 1 - 2\phi\left(\frac{100\epsilon\%}{CV(\hat{N})}\right)$$

when the CV is expressed as a percent. Because $1 - 2\phi(Z) = 0.95$ when $Z = -1.96 \approx -2$, then with 95% certainty, the percentage error in the abundance estimator will not exceed twice the CV of $\hat{N}$.

This same interpretation of the precision of the abundance estimator $\hat{N}$ may be applied to the density estimator $\hat{D} = \hat{N}/A$, because

$$\left|\frac{\hat{N} - N}{N}\right| = \left|\frac{\hat{D} - D}{D}\right|$$

thus extending the utility of $CV(\hat{N})$ as a measure of precision. Another less evident interpretation is available in terms of the percentage error in the reciprocal, $1/\hat{D}$, measuring the average area per individual; thus

$$P\left(\frac{\left|\frac{1}{\hat{D}} - \frac{1}{D}\right|}{\frac{1}{D}} < \epsilon\right) \approx 1 - 2\phi\left(\frac{-\epsilon}{CV(\hat{N})}\right). \qquad (3.35)$$

Indeed, the approximation may be somewhat tighter with respect to the latter interpretation than the former because $1/\hat{N}$ may be more symmetrically distributed about $1/N$ than the skewed sampling distribution of $\hat{N}$ about N.

This skewness in the sampling distribution of $\hat{N} - N$ about zero, however, actually has little effect on the validity of the approximation (3.33), which pertains to the sampling distribution of $|\hat{N} - N|$ rather than $\hat{N} - N$. The skewness acts only to invalidate the equal-tailed interpretation of a percentage error in $\hat{N}$. Whereas there may be a 5% chance that the *absolute* error in $\hat{N}$ will exceed twice the CV, this 5% does not necessarily split equally between over- and underestimation of N. In the case of a Lincoln Index based on small samples, most of this 5% would be due to overestimation by more than two $CV(\hat{N})$. In terms of the "area per animal" interpretation, however, the two tails would more nearly equal 2.5% in probability.

As shown in Figure 3.1, the sampling distribution of $\ln \hat{N}$ for a Lincoln Index is more nearly bell-shaped than the frequency distribution of $\hat{N}$, and this fact may be exploited to obtain improved equal-tail probability bounds on the percentage error in $\hat{N}$. This approximate equality in tail probabilities is achieved through inequality in the upper and lower percentage bounds; for instance

$$P\left(e^{-\epsilon} - 1 < \frac{\hat{N} - N}{N} < e^{\epsilon} - 1\right) = P(|\ln \hat{N} - \ln N| < \epsilon)$$

$$\approx P\left(\frac{-\epsilon}{\sqrt{\text{Var}(\ln \hat{N})}} < Z < \frac{\epsilon}{\sqrt{\text{Var}(\ln \hat{N})}}\right)$$

$$\approx P\left(\frac{-\epsilon}{CV(\hat{N})} < Z < \frac{\epsilon}{CV(\hat{N})}\right)$$

$$= 1 - 2\phi\left(\frac{-\epsilon}{CV(\hat{N})}\right) \qquad (3.36)$$

where $CV(\hat{N})$ is expressed in decimals. This results in yet another useful interpretation of the 95% probability associated with two $CV(\hat{N})$; namely

$$P\left(\frac{\hat{N} - N}{N} < e^{-2CV(\hat{N})} - 1\right) \approx P\left(\frac{\hat{N} - N}{N} > e^{2CV(\hat{N})} - 1\right) \approx 0.025.$$

In some circumstances, there may be reason to express precision in terms of the relative error in $\ln \hat{N}$ because the logarithm of animal density might be interpretable as an index of fitness. For instance, in Chapter 5, treatment conditions will be randomized to replicate study plots of equal size in tests of treatment hypotheses. By this process of randomization, the expected population size at time t_0, $E(N)$, is the same for all study plots at the beginning of the experiment. Consequently, a subsequent mark–recapture survey at time t_1 estimates not only animal abundance but also the ratio of population change through time, e.g., $\hat{N}_1/E(N_0)$. This rate of change can be perceived as an index of fitness of the population under the test conditions. And because only the relative magnitude of such an index is meaningful, then likewise, only the relative magnitude of the estimation error in an index of fitness is relevent. The probability that the relative error in $\ln \hat{N}$ is no more than ϵ is thus approximated by

$$P\left(\left|\frac{\ln \hat{N} - \ln N}{\ln N}\right| < \epsilon\right) \approx 1 - 2\phi\left(\frac{-\epsilon \ln N}{CV(\hat{N})}\right), \qquad (3.37)$$

again depending on the basic measure of precision, $CV(\hat{N})$.

Design of Field Studies

The task of designing mark–recapture surveys was described by Robson and Regier (1964) as a "problem of lifting oneself by one's bootstrap," and this is still true today. A preliminary estimate of population size is necessary before the anticipated precision of an abundance study can be calculated and a sampling design specified. The interesting consequence of this process is that the resulting population surveys ends up as an "objective confirmation or an improvement upon this earlier guess" (Robson and Regier 1964). A preliminary survey, historical data, or an educated guess often serves to provide this initial population figure for sample size calculations.

Other preliminary information needed in the design of field studies is an initial "guesstimate" of the anticipated per-period capture probability to be encountered during the survey. This average capture probability will hopefully be a function of trapping effort, but also will be a function of animal behavior in response to environmental and test conditions. The individual capture probabilities that contribute to this average rate may differ among animals, be influenced by the capture history of individuals, or change over time. However, at the design stage of a field study, the true nature of the variation in capture probabilities is unpredictable and the possibilities are endless. To *a priori* specify a complex structure for distribution of capture probabilities would be unwarranted and possibly misleading. White et al. (1982, p. 167) concluded when calculating sample size, "the only practical approach is to make the best guess at the overall average probability of first capture, $\bar{p}$, applicable during the study...."

The use of best guesses of population size (N) and capture probabilities (p) in sample size calculations is admittedly crude, but as Robson and Regier (1964) state, "it is presently the only approach available to us, and the logic of the situation suggests that no substantially more refined approach is possible." The methodology on preliminary surveys discussed in Chapter 2 may take some of this guesswork out of sample size calculations by providing more accurate estimates of the important design parameters. The time spent on design evaluation will be well spent if "better" strategies can be identified, because the costs of a faulty study are always high.

A prudent investigator may wish to bracket the "best initial guesses" of anticipated population size and capture probabilities with values that represent some of the least and most favorable conditions anticipated. It is not necessary to limit oneself to a single set of design calculations. The cost of performing such calculations will be relatively inexpensive compared to the costs of fieldwork. When the most favorable sampling conditions predict an unacceptable level of precision, a major change in the study design is indicated. Either an increase in the size of the study plot or trapping effort will be necessary to improve precision. A conservative approach to the design of field studies is to anticipate the least favorable field conditions. Under these circumstances, the resulting field study, if different from that anticipated, will provide greater precision than initially deemed necessary.

One additional comment about the use of an average anticipated capture probability in design calculations seems necessary before proceeding. For the single mark–recapture method, Robson and Regier (1964) found that an equal division of sampling effort between marking and recapture is an optimal allocation of resources, assuming equal costs for marking and recapture. Similarly, Wittes (1974) found that the sampling variance of a multiple mark–recapture model [i.e., model M_t (Otis et al. 1978)] is minimized when capture probabilities between trapping periods are equal (i.e., $p_i = p$, for 1, . . . , k). From this perspective, the best strategy in designing an abundance study is to proceed exactly as we must do, to assume a constant rate of capture. However, it also indicates that a deviation from this ideal will result in a survey with potentially less precision than initially intended.

In a single-plot study to estimate animal abundance, there are two possible scenarios that may be encountered in the design of a study. The first of these scenarios to be discussed occurs when a specific population of interest has been identified for study (Figure 3.2). The choice of this population is based on an inherent interest in the abundance of the species at this locality. The study site may be a discrete area within the landscape or simply part of the larger contiguous habitat. With the population size fixed at N by the specification of the study site, the design calculations reduce to finding the capture probability needed to attain a specified level of precision for the abundance estimate ($\hat{N}$).

The second sampling scenario that may be encountered in a mark–recapture study is the case where abundance is to be estimated at a plot chosen as representative of the larger contiguous landscape (Figure 3.2). As

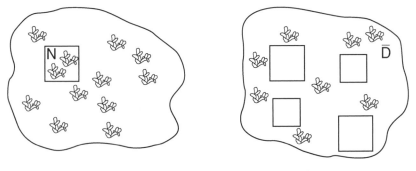

Case 1   Case 2

Figure 3.2 Scenarios that can be encountered in designing field studies to estimate animal abundance: case 1, where the location and site of the population (N) is fixed, and case 2, where the location of the site is yet undetermined and only the mean density ($\bar{D}$) anticipated.

such, both the size of the study plot with its associated animal abundance and the level of capture probability are unspecified. Consequently, the parameters N and p are selected jointly to yield an abundance estimate with prescribed precision. To perform the sample size calculations, an initial estimate of the mean density $(\bar{D})$ of the wild population is needed. Then, by finding the necessary level of population abundance $(\hat{N})$, plot size (A) is determined from the relationship $A = N/\bar{D}$.

An interesting aspect of this second scenario for abundance studies is that the population size at the study plot is now a random variable. The actual abundance at the site will be something greater than or less than the population level predicted by the mean density of the species. Despite the uncertainty of the number of animals on the site, the sample size calculations presented above still pertain and provide a realistic estimate of the required capture rate. For instance, using the expression (3.33), the variance of the objective function is approximately

$$\text{Var}_N\left[\frac{\hat{N} - N}{N}\right] = E_N\left[\text{Var}\left(\frac{\hat{N} - N}{N}\bigg|N\right)\right] + \text{Var}_N\left[E\left(\frac{\hat{N} - N}{N}\bigg|N\right)\right] \approx E_N\left[\frac{\text{Var}(\hat{N})}{N^2}\right]$$

and in the case of the Lincoln Index,

$$\text{Var}_N\left[\frac{\hat{N} - N}{N}\right] \approx \frac{q_1 q_2}{\mu_N p_1 p_2}. \tag{3.38}$$

In expression (3.38), the μ_N denotes the mean animal abundance across study plots. Hence, the precision of the abundance estimate depends only on the sampling error of the mark–recapture study (3.38). The reason is that the component for the spatial variance in abundance is eliminated when the difference between N and $\hat{N}$ is taken in the expression for precision (3.33). The consequence is that an abundance study can be effectively designed, whether a specific population has been identified for study or simply a representative plot is to be chosen from a segment of the range of the species.

Lincoln Index   The sample size calculations will be illustrated first for the Lincoln Index and then for the constant probability, multiple-recapture, and removal sampling techniques. Using the definition of precision given by (3.33) and using the fact that $\ln\hat{N}$ for a Lincoln Index is approximately a normal random variate, the necessary per-period capture probability is calculated as follows:

$$P\left(\left|\frac{\hat{N} - N}{N}\right| < \epsilon\right) = 1 - \alpha$$

$$P(N - \epsilon N < \hat{N} < N + \epsilon N) = 1 - \alpha$$

$$P(\ln N(1 - \epsilon) - \ln N < \ln \hat{N} - \ln N < \ln N(1 + \epsilon) - \ln N) = 1 - \alpha$$

$$P\left(\frac{\ln(1 - \epsilon)}{\sqrt{\text{Var}(\ln \hat{N})}} < Z < \frac{\ln N(1 + \epsilon)}{\sqrt{\text{Var}(\ln \hat{N})}}\right) = 1 - \alpha$$

$$P\left(Z < \frac{N \ln(1 + \epsilon)}{\sqrt{\text{Var}(\hat{N})}}\right) - P\left(Z < \frac{N \ln(1 - \epsilon)}{\sqrt{\text{Var}(\hat{N})}}\right) = 1 - \alpha$$

$$\phi\left(\frac{\sqrt{N} \ln(1 + \epsilon)}{(1/p) - 1}\right) - \phi\left(\frac{\sqrt{N} \ln(1 - \epsilon)}{(1/p) - 1}\right) = 1 - \alpha$$

$$(3.39)$$

where again ϕ represents the cumulative standard normal distribution. For fixed N, the Newton–Raphson method (Seber 1982, pp. 16–18; Press et al. 1986, pp. 240–273) can be used to iteratively solve for p when ϵ and α have been specified. At $\alpha = 0.10$, Eq. (3.39) has been solved in terms of N and p (Figure 3.3) for various values of $\epsilon(0.01 - 0.90)$. For example, with a population of size $N = 100$, a capture probability of $p = 0.45$ is necessary for an abundance study to have an anticipated precision given by $\epsilon = 0.2$, $\alpha = 0.10$. In other words, with $p = 0.45$ and $N = 100$, $\hat{N}$ can be expected to be within $\pm 20\%$ of the value of N, 90% of the time.

The iterative nature of solutions to Eq. (3.39) suggests the need for a more convenient means of calculating sample size in Lincoln Indices. Defining precision by the expression

$$P\left(\left|\ln\left(\frac{\hat{N}}{N}\right)\right| < \epsilon\right) = 1 - \alpha$$

and once again assuming $\ln \hat{N}$ approximately distributed as a normal random variate, the necessary capture probability has the closed solution

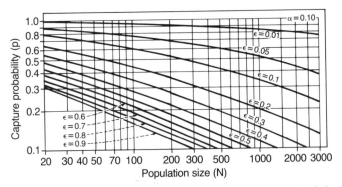

Figure 3.3 The per-period capture probability (p) and population size (N) needed in a Lincoln Index study for a level of precision defined by $\alpha = 0.10$ and $\epsilon = 0.10 - 0.90$. Sample size calculations are based on (3.39).

$$p = \cfrac{1}{\left[1 + \cfrac{\epsilon \sqrt{N}}{Z_{1-(\alpha/2)}}\right]} \qquad (3.40)$$

For example, suppose the best guess for the abundance of Townsend's ground squirrel (*Spermophilus townsendis*) on a 6.5-ha study plot is $N = 200$ and one wishes the subsequent abundance estimate to have a precision of $P(0.8 < \hat{N}/N < 1.25) = 0.95$ (i.e., $\epsilon = 0.223$, $\alpha = 0.05$); then from (3.40), the necessary capture probability per period is found to be $p = 0.383$. Alternatively, assuming that the numbers of animals captured per period can be predetermined in advance, Robson and Regier (1964) give sample sizes for a Lincoln Index for precision specified by $(1 - \alpha) = 0.95$ and $\epsilon = 0.1$, 0.25 and 0.5.

Multiple-Mark–Recapture   Lee (1972), by assuming the number of distinct animals captured (r) in a k-sample multiple-mark–recapture survey to be normally distributed, developed curves for sample size calculations for the model M_0 (Figure 3.4). Defining precision by (3.33), Lee (1972) developed a total of 48 graphs where the average capture probability (p) necessary per trapping period was plotted as a function of population size N and the number of sampling periods (k). Parts of three of the graphs $[(1 - \alpha) = 0.90$ and $\epsilon = 0.1$, 0.2, and 0.50] have been reproduced (Figure 3.4).

Removal Sampling   For removal sampling, no comparable graphs or tables as found in Robson and Regier (1964) and Lee (1972) have been computed to assist in sample size calculations. Zippin (1958) gave a small table that indicated the fraction of a population that must be removed to achieve a specified *CV*. The table considered only six population levels and consequently has little practical use. Based on the assumption that $1/\hat{N}$ for removal sampling is approximately normally distributed (Skalski et al. 1984), the value of p necessary for a specified level of precision can be determined. Using (3.33) once again as the precision function, we can find levels of N and p by iteratively solving the equation

$$\phi\left(\frac{N\epsilon}{(1 - \epsilon)\sqrt{\text{Var}(\hat{N})}}\right) - \phi\left(\frac{-N\epsilon}{(1 + \epsilon)\sqrt{\text{Var}(\hat{N})}}\right) = 1 - \alpha \qquad (3.41)$$

for specified values of ϵ and α with $\text{Var}(\hat{N})$ given by (3.18). Because numerical methods must be used to solve Eq. (3.41), graphs have been developed for the specification $(1 - \alpha) = 0.90$ and $\epsilon = 0.1$, 0.2 and 0.5 (Figure 3.5). For populations estimated by change-in-ratio methods, Paulik and Robson (1969) provide curves to determine sample size for specified precision (3.33).

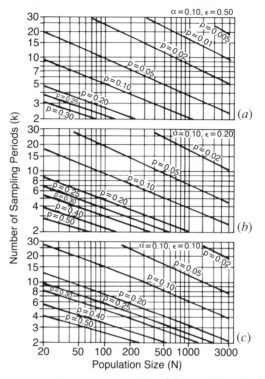

Figure 3.4 The per-period capture probability (p), population size (N), and number of sampling periods (k) needed for a multiple-mark–recapture study when precision is defined as (3.33) and where $\alpha = 0.10$ and (a) $\epsilon = 0.10$, (b) $\epsilon = 0.20$, (c) $\epsilon = 0.50$. Figures are from Lee (1972).

The distinction between the two sampling scenarios (Figure 3.2) for abundance studies can be seen in the way sample size charts (Figures 3.3–3.5) are used. In the first scenario, where population size is predetermined by the specification of the study area, one simply identifies the value of p that intersects with the value of N along the chosen precision contour given by α and ϵ. In the second scenario, any point along the chosen precision contour can be selected as the basis for the design of the abundance study. Within the constraints of the sampling model, all points along a contour can be expected to provide the same level of precision. We will investigate the second scenario in the next section to determine which combination of N and p is optimal for fixed cost.

The sample size calculations thus far have resulted in specifying levels of capture probability (p) necessary for successful abundance estimation. To

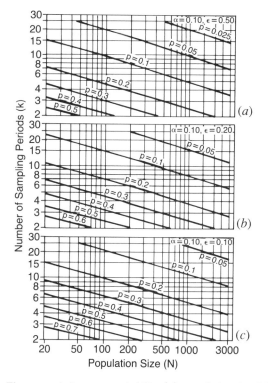

Figure 3.5 The per-period capture probability (p), population size (N), and number of sampling periods (k) needed for a removal study when precision is defined as (3.33) and where $\alpha = 0.10$ and (a) $\epsilon = 0.10$, (b) $\epsilon = 0.20$, (c) $\epsilon = 0.50$. Sample size calculations are based on (3.41).

be useful, these values of p must be further translated in terms of trapping effort. A catch–effort relationship expresses the numerical relationship between the number of traps employed (f) and the probability of capturing an animal (p), in general

$$g(f) = p. \tag{3.42}$$

In Chapter 2, the catch–effort model $p_i = 1 - e^{-cf/A}$ was proposed as a possible relationship for predicting the probability of capture from the level of trapping effort, where c is an estimable catch coefficient and A is plot size. And as such, $g(f)$ is used to determine the trapping effort necessary to attain a specified level of precision in abundance studies.

Using the closed form of sample size calculations derived for the Lincoln Index, Eq. (3.40) translates to

$$g(f) = \cfrac{1}{\left[1 + \cfrac{\epsilon \sqrt{N}}{z_{1-(\alpha/2)}}\right]}$$

and using $g(f) = 1 - e^{-cf/A}$ yields

$$f = -\frac{A}{c} \ln \left[1 - \cfrac{1}{\left[1 + \cfrac{\epsilon \sqrt{N}}{z_{1-(\alpha/2)}}\right]}\right]. \qquad (3.43)$$

Trapping effort at a study plot is now expressed as a function of the desired precision of the abundance study. In the Townsend's ground squirrel example, a catch coefficient of $c = 0.0325$ (i.e., $c/A = 0.005$) would indicate a need for 97 traps distributed across the 6.5-ha study area to achieve the $p = 0.383$ capture probability deemed necessary for the precision desired ($\epsilon = 0.223$, $\alpha = 0.05$).

Design Optimization

The purpose of design optimization attempts to go beyond simply identifying a successful design, for it seeks the most economical design for the stated study objective. The purpose of design optimization is to devise a field study that most efficiently allocates sampling efforts consistent with our knowledge of the states of nature, economic resources, and sampling objectives. To accomplish this task, prior estimates of various parameters must be available and design relationships predicted in order that the optimal allocation scheme can be identified (see Chapter 2). Fortunately, optimization procedures are rather robust, permitting even crude estimates of parameters and interrelationships to provide results near the optimum (Hansen et al. 1953, p. 284).

Two approaches to the optimization of abundance estimation have been used in the design of field studies. One approach is to consider the population of interest as predetermined and of fixed size N (case 1, Figure 3.2) (Robson and Regier 1964), while the other approach (Skalski 1985b) assumes that both sampling effort and population size are variables to optimally determine (case 2, Figure 3.2). The proper approach will depend on the situation an investigator encounters.

In Robson and Regier (1964), a cost function of the form

$$C_0 = C_1 n_1 + C_2 n_2 \qquad (3.44)$$

where C_0 = total variable costs of conducting the study, C_1 = cost of capturing and marking an animal, and C_2 = costs of capturing and examining an animal, was assumed in optimizing the numbers of animals captured, n_1 and n_2, during the respective sampling periods of a Lincoln Index. The optimal numbers of animals to capture are those values of n_1 and n_2 that satisfied the equation

$$\frac{n_1 C_1}{n_2 C_2} = \frac{N - n_2}{N - n_1} \tag{3.45}$$

for fixed total costs C_0. By simultaneously solving the two equations, (3.44) and (3.45), the two unknowns, n_1 and n_2, can be easily found. This specification of an optimal design requires information on costs and an initial estimate of the size of the fixed population (N).

In Skalski (1985b), a cost function of the form

$$C_0 = C_1 f + C_2 \sqrt{Af} \tag{3.46}$$

is used in conjunction with the expression for the CV of a density estimate $(\hat{D} = \hat{N}/A)$ from a Lincoln Index,

$$CV(\hat{D}) = \sqrt{\frac{1}{DA}\left(\frac{1}{p} - 1\right)^2} \times 100\%$$

to optimize the design of a single-plot study. Using the catch–effort model

$$p = 1 - e^{-cf/A}, \tag{3.47}$$

the optimal plot size is shown to be $A_{opt} = 0$ with trapping effort of $f_{opt} = C_0/C_1$ when the objective is simply to maximize the precision of the subsequent abundance estimate. The reason for this anomaly is that by simply focusing on precision, a plot size can be chosen small enough that an entire population can be almost assured of complete enumeration (i.e., as $A \to 0$, $N \to 0$ while $f \to \infty$), thereby driving the sampling error to zero. The problem with this approach is that as plot size becomes increasingly small, the plot can no longer be considered representative of the larger contiguous population from which it was selected. Consequently, the objective of an abundance study (case 2, Figure 3.2) should be to maximize both the precision and the representativeness of the resulting abundance estimate.

To optimize both the representativeness and the precision of an abundance study, an investigator must consider, in addition to sampling error, the variance in N across all possible study plot locations. The unconditional variance in this case is then expressed as

$$\text{Var}(\hat{N}) = \text{Var}_N[E(\hat{N} \mid N)] + E_N[\text{Var}(\hat{N} \mid N)]$$

$$= \sigma_N^2 + \frac{\mu_N q_1 q_2}{p_1 p_2} \tag{3.48}$$

for a Lincoln Index. This unconditional variance expression (3.48) consists of not only the sampling error about the survey estimate $(\hat{N})$ but also spatial variance in animal abundance (σ_N^2) across the landscape. In this context, σ_N^2 provides a measurement of the uniformity of the animal abundance from site to site. The variance component, σ_N^2, will increase as the animal dispersion becomes more clumped. Thus, optimizing the representativeness and the precision of a characteristic plot study is equivalent to minimizing (3.48) within the economic constraints imposed by cost Eq. (3.46).

In order to perform this optimization, a mathematical relationship must first be established between the anticipated magnitude of σ_N^2 and plot size (A). Considering mean–variance relationships exhibited by various models for dispersion and noting stochastic processes that have been used to generate dispersion patterns, the following generalization of (2.26) is adopted:

$$\sigma_D^2 = \left(\frac{D}{A}\right)\eta + \omega D^2. \tag{3.49}$$

For instance, if animal abundance is randomly distributed (i.e., Poisson), then $\eta = 1$ and $\omega = 0$ such that

$$\sigma_D^2 = \frac{D}{A}.$$

The unconditional CV of $\hat{D}$ for a Lincoln Index using variance relationships (3.48) and (3.49) can be expressed as

$$CV(\hat{D}) = \sqrt{\frac{1}{DA}[\eta + (e^{cf/A} - 1)^{-2}] + \omega} \times 100\%. \tag{3.50}$$

Solving the CV (decimal) in terms of f, we obtain

$$f = \frac{A}{c} \ln\left[1 + \frac{1}{\sqrt{DA(CV(\hat{D}))^2 - \omega AD - \eta}}\right]. \tag{3.51}$$

The optimal design configuration for a single-plot study can be found through a process of trial and error using Eqs. (3.46) and (3.51).

To optimize a study design for fixed costs, C_0, the preceding analysis is reversed. First, the cost contour (3.46) is solved and plotted as a function of f and A. Next, the precision function (3.51) is solved and plotted to find the minimum value of $CV(\hat{D})$ whose curve is tangential to the cost contour previously drawn. The intersect point identifies the combination of f and A that maximizes the precision of the field study for fixed budget, C_0. Furthermore, a series of such solutions can be used to plot a third curve, the optimal design contour (dotted line, Figure 3.6). Once established, an optimum for either fixed cost or precision can be identified from the intersection of the optimal design curve with either the cost contour or precision contour, respectively.

Using a Taylor series approximation (i.e., $\ln(1 + x) \approx x$ for $x \le .3$) to Eq. (3.51),

$$f \approx \frac{A}{c\sqrt{DA\,(CV\,(\hat{D}))^2 - \omega AD} - \eta} \tag{3.52}$$

permitting a closed solution to the optimization of a single-plot study for fixed precision. The optimal design configuration can be found by noting that the optimum using Eq. (3.51) occurs at or to the left of its minimum. Thus, the value of A which minimizes (3.52) and provides an approximate solution to the optimization problem is

$$A_{\text{opt}} = \frac{2\eta}{D\,(CV\,(\hat{D}))^2 - \omega D} \tag{3.53}$$

with the corresponding trap effort of

$$f_{\text{opt}} = \frac{A_{\text{opt}}}{c\sqrt{\eta}}. \tag{3.54}$$

The reason for using the Taylor series approximation (3.52) is in the insights results (3.53) and (3.54) provide. For the special case of Poisson-distributed abundance ($\eta = 1$, $\omega = 0$), the optimal plot size becomes

$$A_{\text{opt}} = \frac{2}{D \cdot CV\,(\hat{D})^2} \tag{3.55}$$

with an optimal trapping effort of

$$f_{\text{opt}} = \frac{A_{\text{opt}}}{c}. \tag{3.56}$$

and a corresponding trap density (i.e., number of traps/unit area) of

$$\frac{f_{\text{opt}}}{A_{\text{opt}}} = \frac{1}{c}, \tag{3.57}$$

the reciprocal of the Poisson catch coefficient. The importance of a catch effort model and a prior estimate of the catch coefficient is now evident.

EXAMPLE A graphical approach to optimization of abundance studies makes finding the design parameters f and A relatively easy. For example, assume that survey data suggest the variance–plot size relationship

$$\sigma_{\hat{D}}^2 = \frac{35}{A} + 2$$

for a population of least chipmunks (*Eutamias minimus*) with an average density of 10 animals per hectare (i.e., $\eta = 3.5$, $\omega = 0.02$, $D = 10$). Further, assume that the catch–effort relationship (3.47) is applicable with a catch coefficient of $c = 0.01$. Using these parameter esti-

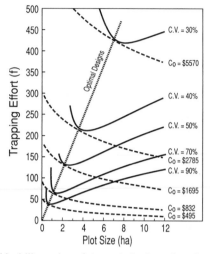

Figure 3.6 Graphical illustration of the optimization of an abundance study whose objective is population estimation. Cost contours [dashed lines, (3.58)], precision contours [solid lines, (3.51)] and optimal design contour [dotted line] illustrate solutions to the optimal design of a study of the abundance of the least chipmunks (*Eutamias minimus*) with an anticipated density of 10 ha^{-1} and catch coefficient $c = 0.01$.

mates for animal dispersion and catchability, Eq. (3.51) can be solved in terms of f and A for a prespecified level of $CV(\hat{D})$ [e.g., $CV(\hat{D}) = 50\%$; solid line, Figure 3.6]. The next step is to solve the cost function (3.46) in terms of various combinations of f and A and plot these solutions on the same graph. For the cost function

$$C_0 = \$8.33f + \$36.49\sqrt{Af} \tag{3.58}$$

pairwise values of f and A can be plotted (dashed line, Figure 3.6) for different choice of budget, C_0. The value of C_0 whose cost contour shares a common tangent with the precision contour of prespecified $CV(\hat{D})$, identifies the optimal design for fixed precision. The intersect point gives the values of f and A that minimize cost, C_0, for fixed precision $CV(\hat{D})$. With a $CV(\hat{D}) = 50\%$, the minimum cost of a field study providing that level of precision is $C_0 = \$1,695$ based on $A = 2.2$ ha and $f = 127$ traps. Any other field design providing the same level of precision will necessarily cost more to perform. The Taylor series approximation [$\ln(1 + .8006) = .5881$] is not sufficiently precise to permit the use of the closed solutions (3.53) and (3.54) for this example, which yield values of $A_{opt} = 3.04$ and $f_{opt} = 162$, respectively.

The exercise of design optimization, as illustrated above, must be viewed realistically. It is a tool that can be abused as can any other useful device. First, the results of the calculations are specific to only the census technique used in the computations. Other models, such as those of multiple-mark–recapture or removal sampling, can be evaluated in a similar manner. Circumstances then may allow the selection of the census technique with the best predicted performance under the economic constraints of the study. However, optimization techniques do not take into account the robustness of certain field sampling designs to violations in model assumptions. Therefore, results of optimization must be viewed in terms of the unpredictable and varying nature of trapping studies.

It is the responsibility of an investigator to choose a sampling technique that will be robust to violations in model assumptions. In designing a population survey, the possibility that one or more of the assumptions of the simple models used in sample size calculations may be violated must be taken into account. For instance, the Lincoln Index remains valid despite various violations in the assumption of closure and requires only one of its two sampling periods to be a random sample of the population. Alternatively, the multinomial models of Otis et al. (1978) of which models M_0 and M_B are special cases, allow for limited trap response and heterogeneity in individual capture probabilities. We recommend investigators perform similar optimality calculations for these and other survey models they may use.

Optimal designs for abundance studies [(3.53) and (3.54) or (3.55) and (3.56)] can be expected to perform suboptimally when model violations occur. There is no reason to believe, however, that a suboptimal design will respond any better to model violations than will a design that intentionally tries to maximize the performance of a field study. This is because an optimal sampling design can be expected to maximize the numbers of animals captured, which, in turn, is exactly what is needed if model violations are to be detected and appropriate model adjustments made.

Recommendations

The literature associated with abundance estimation typically has focused on the statistical theory of parameter estimation. Additionally, studies that have attempted to compare the performance of estimation techniques under field conditions have largely provided insights based on a single trial and are species-specific (Edwards and Eberhardt 1967; Smith 1968). Together, the available literature has left researchers to their own devices and personal experience in the design and analysis of population studies. The best prospects for improving the performance of abundance studies is to couple the intuition and experience of dealing with an animal species and the quantitative guidance available from the knowledge of the statistical behavior of population estimators. The discussions above permit some general guidelines on the design and analysis of abundance studies to be stated.

Recommendations for the design and analysis of population surveys include the following:

1. Recognize the optimal design of a field investigation changes with the objective of the study (e.g., a manipulative experiment is not simply a series of interrelated population surveys).

2. Specify the objective of the study and subsequent data analysis at the time the study is being designed to assure that the purpose of the investigation is simply population estimation.

3. Incorporate spatial or temporal variance components into design considerations if inferences to a wider collection of study plots, sampling periods, or populations are intended.

4. Specify an objective function for the investigation such as the sampling precision of an abundance estimate or population contrast.

5. Use prior survey data or site reconnaissance to obtain "guesstimates" of the likely abundance levels (N) and average per-period capture probabilities ($\bar{p}$) to be encountered during the course of the field investigations.

6. Perform sample size calculations based on such models as the Lincoln Index, constant probability removal model or multiple-recapture model (M_0), which assume a simple structure for trap response.

7. Repeat sample size calculations under a variety of likely scenarios for the possible outcome of a field investigation (i.e., determine the projected performance of a study in the presence of errors in prior guesstimates of N and $\bar{p}$).

8. Implement a survey design that permits model selection in the choice of the most appropriate abundance estimator.

These recommendations attempt to minimize the influence of ill-conceived study objectives and the role of chance on study performance. To do this, the methods presented in this chapter have focused on characterizing the stochastic nature of capture data in order to improve the design the abundance studies. In Chapters 5 and 6, the stochastic nature of the wild populations will be included with these considerations for the purposes of designing and analyzing manipulative experiments and assessment studies.

4

Comparative Censuses

The simplest form of a comparative wildlife study is the quantitative comparison of animal abundance between two temporally or spatially distinct populations, termed a "comparative census" by Skalski et al. (1983a, 1984). Commonly posed as a test of the hypothesis (Chapman 1951; Chapman and Overton 1966; Seber 1982, p. 121)

$$H_0: N_1 = N_2 \qquad\qquad (4.1)$$
$$H_a: N_1 \neq N_2$$

where $N_i (i = 1, 2)$ is the animal abundance in the ith population, it shares some of the same conceptual difficulties as hypothesis (3.29). Only by an accident of chance will the null hypothesis (4.1) ever be true. Hence, a test of such a null hypothesis is statistically meaningless and biologically uninformative. The outcome of such a test will depend solely on the precision of the surveys of population abundance relative to the actual size of the difference that exists between the two populations. With sufficient sampling effort (allowing capture probabilities to approach 1.0), the null hypothesis (4.1) will be rejected with absolute certainty. Whenever a test is guaranteed to reject a null hypothesis when sufficient sampling effort is used, it inevitably indicates a poorly conceived hypothesis.

Comparative censuses prompt for the first time a need to consider the nature of the population response being tested. A hypothesis of form (4.1) is comparing the abundance of two populations, but it is also describing the effect of differences in locality or time on animal abundance. If the purpose of the surveys is to estimate the difference in sizes of the two populations, the numerical difference, $\hat{N}_2 - \hat{N}_1$, is always appropriate. In contrast, if the comparison is to be descriptive of an effect on animal abundance, the nature of the comparison is important.

Consider a model equation that describes animal abundance at a study plot as a linear function, for example

$$\hat{N}_i = \mu + \tau_i + \epsilon_i \qquad (4.2)$$

where $\hat{N}_i$ = estimate of animal abundance (N) at the ith plot

μ = overall mean abundance on the study plots

τ_i = effect of location $i(i = 1, 2, \ldots)$ on realized abundance

ϵ_i = random-error term associated with the stochastic variation in animal abundance across the landscape $(i = 1, 2, \ldots)$ and measurement error

In Eq. (4.2), plot location is assumed to have an additive effect on animal abundance. A numerical difference in population levels assuming response model (4.2)

$$\hat{N}_2 - \hat{N}_1 = (\mu + \tau_2 + \epsilon_2) - (\mu + \tau_1 + \epsilon_1) = \tau_2 - \tau_1 + \epsilon'$$

provides an uncluttered estimate of the effect of location on animal abundance. This estimate of the location effect $(\hat{N}_2 - \hat{N}_1)$ is a function of the difference in location parameters $(\tau_2 - \tau_1)$ plus a random-error term (ϵ').

Alternatively, the response model could be conceptualized as a product of multiplicative effects, where the abundance estimate is now modeled as

$$\hat{N}_i = \mu \tau_i \epsilon_i \qquad (4.3)$$

Under this multiplicative model, an estimate of location effects is now estimated from a ratio of abundance estimates where

$$\frac{\hat{N}_2}{\hat{N}_1} = \frac{\mu \tau_2 \epsilon_2}{\mu \tau_1 \epsilon_1} = \frac{\tau_2}{\tau_1} \epsilon'.$$

Consequently, if the purpose of the surveys is to provide an estimate of the effect of location, the nature of the response model and the contrast estimated need to be considered. For example, assuming model (4.3) and calculating an arithmetic difference, the resulting comparison

$$\hat{N}_2 - \hat{N}_1 = \mu \tau_2 \epsilon_2 - \mu \tau_1 \epsilon_1 = \mu(\tau_2 \epsilon_2 - \tau_1 \epsilon_1)$$

does little to clarify the effects of plot location for they remain hopelessly cluttered with other model parameters.

When discussing response models for population studies, the comment of Cormack (1968, p. 456) on the oversimplicity of even the best population models must be recalled. However, in terms of the processes of natality and mortality, a multiplicative model such as (4.3) often may be a more reasonable approximation. A multiplicative effect is assumed to result in a fractional change in the abundance of a population. For example, the effect of secondary poisoning from pesticide sprays may be expected to affect a fixed fraction of the susceptible individuals of a population regardless of abundance level. It is much less likely that a fixed number of animals would succumb to poisoning if a series of populations were so treated. Many other natural and human-induced effects may be expected to behave in a similar multiplicative manner.

Analysis of Comparative Census Data

Estimation of Proportional Abundance
Analysis of a comparative census will be presented in terms of an estimate of proportional abundance. The proportional abundance of one population, N_2, relative to that of another, N_1, is defined as the ratio

$$K = \frac{N_2}{N_1}. \tag{4.4}$$

Skalski et al. (1983a, 1984) found that there are basically two means of estimating K from capture data. One method is based on estimates of absolute abundance ($\hat{N}$) and the other, on the use of catch indices (r). The choice depends on whether the capture probabilities are homogeneous between the populations being compared. When capture rates are homogenous between populations, proportional abundance estimator $\hat{K} = r_2/r_1$ will provide a valid estimate with a smaller sampling variance than $\hat{K}$ based on absolute estimates. Tests of homogeneity presented in Chapter 3 should be used to determine whether absolute abundance or catch indices should be used in estimating (4.4).

Lincoln Indices When the two populations of a comparative census are sampled under the assumptions of the single-mark–recapture method (Skalski et al. 1983a), a conditionally unbiased (for $n_{21} + n_{22} > N_2$) estimator of K is

$$\hat{K}_1 = \frac{\dfrac{(n_{21} + 1)(n_{22} + 1)}{(m_2 + 1)} - 1}{\dfrac{n_{11} n_{12}}{m_1}} \tag{4.5}$$

with

$$\text{V\^ar}(\hat{K}_1 \mid N_1, N_2) = \frac{m_1 n_{21} n_{22}}{m_2^3 n_{11}^3 n_{12}^3}[(n_{22} - m_2)(n_{21} - m_2)m_1 n_{11} n_{12}$$

$$+ (n_{12} - m_1)(n_{11} - m_1)m_2 n_{21} n_{22}] \tag{4.6}$$

where n_{i1} = number of animals marked in the initial sampling period for
the ith population (i = 1, 2)

n_{i2} = number of animals captured in the second period of the
survey for the ith population (i = 1, 2)

m_i = number of marked and recaptured animals in the ith
population (i = 1, 2)

Alternatively, when the capture probabilities can be assumed to be homogeneous between populations, proportional abundance can be estimated by

$$\hat{K}_2 = \frac{r_2}{r_1} \tag{4.7}$$

with

$$\text{V̂ar}(\hat{K}_2 \mid K) = \frac{(n_{.2} - m_.)(n_{.1} - m_.)(\hat{K}_2 + \hat{K}_2^2)}{n_{.1} n_{.2} r_1} \tag{4.8}$$

for $r_i = n_{i1} + n_{i2} - m_i$ and where the dot ($\cdot$) denotes summation across populations. A $(1 - \alpha)100\%$ confidence interval estimate for K using either (4.5) or (4.7) can then be computed from the algorithm

$$\left[\hat{K} \exp \pm Z_{1-\alpha/2} \sqrt{\frac{\text{V̂ar}(\hat{K})}{\hat{K}^2}} \right] \tag{4.9}$$

where ln $\hat{K}$ for a Lincoln Index is approximately normally distributed. The computations for a comparative census are illustrated in Skalski et al. (1983a) using data from two surveys of a Nuttall's cottontail (*Sylvilagus nuttallii*) population.

Multiple-Mark–Recapture Models   Analysis of comparative census data based on multiple-mark–recapture data begins with a test of interpopulation homogeneity (3.15). The test of homogeneity is valid regardless of the survey model(s) ultimately fit to the mark–recapture data using either program CAPTURE (Otis et al. 1978) or one of the closed-population models in Cormack (1985). If homogeneity is indicated, proportional abundance should be estimated using (4.7). A conclusion of homogeneity also implies that the two populations share a common fit to a survey model. The variance estimate for (4.7) can then be written as

$$\text{V̂ar}(\hat{K}_2 \mid K) = \frac{(1 - \hat{P})}{r_1}(\hat{K}_2 + \hat{K}_2^2) \tag{4.10}$$

where $\hat{P} = (1 - \Pi_{i=1}^k (1 - \hat{p}_i))$ is the estimated overall probability of capture during the k-period multiple-mark–recapture survey, and where $\hat{p}_i$ is the estimated probability of capture in the ith period ($i = 1, \ldots, k$) from the survey model fit to the pooled capture data.

If interpopulation heterogeneity is detected, proportional abundance should be estimated by

$$\hat{K}_1 = \frac{\hat{N}_2}{\hat{N}_1} \tag{4.11}$$

where the $\hat{N}_i (i = 1, 2)$ are independent estimates from the two separate surveys. During model selection, different survey models might be found appropriate in estimating $\hat{N}_1$ and $\hat{N}_2$. Valid abundance estimation should take precedence over attempting to fit a common survey model to both population surveys when using the estimator (4.11). The estimated variance of $\hat{K}_1$ assuming independent estimates of population abundance is then

$$\hat{\text{Var}}(\hat{K}_1 \mid K) = \frac{\hat{\text{Var}}(\hat{N}_2)}{\hat{N}_1^2} + \frac{\hat{\text{Var}}(\hat{N}_1)\hat{N}_2^2}{\hat{N}_1^4}$$

$$= \hat{K}_1^2 \left(\frac{\hat{\text{Var}}(\hat{N}_1)}{\hat{N}_1^2} + \frac{\hat{\text{Var}}(\hat{N}_2)}{\hat{N}_2^2} \right). \tag{4.12}$$

The distributional behavior of $\hat{K}_1$ and $\hat{K}_2$ may be expected to vary under the wide variety of possible multiple-mark–recapture survey models that may be fit to the data. Without model-specific information, a ln-transformation of the point estimates of K (i.e., $\hat{K}_1$ or $\hat{K}_2$) is recommended in conjunction with confidence interval estimation (4.9).

EXAMPLE   To illustrate the use of the multiple-mark–recapture data in a comparative census, cotton rat (*Sigmodon hispidus*) data from White et al. (1982, pp. 135–149) will be analyzed. The study consisted of 8 consecutive days of trapping at a single site with the objective of the analysis to compare the relative sizes of the male and female populations. Because the two populations were trapped simultaneously with equal effort, the prospect of homogeneous capture probabilities is likely. A contingency table (3.15) analysis (Table 4.1) based on the frequency of the number of animals captured 1, 2, 3, 4, and 5–8 times for the two sexes is

Table 4.1

Contingency Table Analysis of the Numbers of Male and Female
Cotton Rats Captured 1, 2, 3, 4, and 5–8 Times during a
Multiple-Mark–Recapture Survey (White et Al. 1982)

Frequency of capture	Males	Females	Total
1	66	43	109
2	27	20	47
3	11	15	26
4	3	2	5
5–8	1	6	7
Total	108	86	194

only marginally significant $[P(\chi_4^2 \geq 7.8892) = .0957]$ at $\alpha = 0.10$. White et al. (1982) show the pooled data and that of the separate sexes to have an excellent fit to the heterogeneity model M_h [see Otis et al. (1978) and White et al. (1982) for a discussion of the jackknife abundance estimator M_h]. The combined data resulted in an estimate of $\hat{N} = 391$ with $\text{Var}(\hat{N}) = 1294.618$. The male population was estimated be $\hat{N}_M = 211$ with $\text{Var}(\hat{N}_M) = 435.594$, and the female component of the population was estimated to be $\hat{N}_F = 148$ with $\text{Var}(\hat{N}_F) = 231.302$. Rather than performing a test of equality of abundance as was done by White et al. (1982), the sex ratio will be estimated.

$$\hat{K}_1 = \frac{\hat{N}_M}{\hat{N}_F} = \frac{211}{148} = 1.4257$$

with associated variance

$$\text{Var}(\hat{K}_1 \mid K) = (1.4257)^2 \left[\frac{231.302}{(148)^2} + \frac{435.594}{(211)^2} \right] = 0.04135.$$

The resulting 95% confidence interval estimate for the sex ratio of the cotton rat population is computed using (4.9) to be

$$CI(1.4257e \pm [1.96\sqrt{0.04135/(1.4257)^2}]) = 0.95$$

or

$$CI(1.078 < K < 1.886) = 0.95.$$

In other words, one is 95% certain that the number of male cotton rats outnumber females by 7.8% to 88.6%.

Had homogeneity of capture probabilities been plausible, proportional abundance would have been estimated by $\hat{K}_2 = r_M/r_F = 108/86 = 1.12558$. The variance estimator (4.10) would have been based on using the estimate of the overall probability of capture $\hat{P} = (108 + 86)/391 = 0.4962$ [i.e., $\hat{P} = (r_M + r_F)/\hat{N}$], resulting in

$$\text{Var}(\hat{K}_2 \mid K) = \frac{(1 - 0.4962)}{86}(1.2558 + (1.2558)^2) = 0.0166.$$

However, derivation of variance formula (4.10) is based on binomial sampling error and the assumption all animals have equal probability of capture. In the case where differences in capture probabilities exist (i.e., intrapopulation heterogeneity) among individual animals, the binomial sampling variance is too large (Feller 1971, pp. 230–231), and (4.10) will tend to overestimate the true variance of $\hat{K}_2$. A resulting confidence interval using (4.9) will therefore tend to be valid but conservative (i.e., too wide). In general, for any of the multiple-mark–recapture models discussed in Otis et al. (1978) proportional abundance estimator (4.11) and its variance (4.12) is appropriate. Unfortunately, absolute abundance estimators do not exist for all eight models, and so $\hat{K}_1$ cannot be computed. Alternatively, proportional abundance estimator (4.7) is applicable to any set of surveys where interpopulation homogeneity exists. The variance estimator (4.10) is then appropriate under models M_0, M_t, and M_b. However, in the cases of models M_h, M_{th}, M_{tb}, or M_{tbh} where intrapopulation heterogeneity is present, variance formula (4.10) based on binomial sampling will be inappropriate, and no variance formula is anticipated.

Constant Probability Removal Model   When the populations of a comparative census are surveyed using a k-sample removal procedure (Zippin 1956, 1958), proportional abundance can be estimated as

$$\hat{K}_1 = \frac{\hat{N}_2}{\hat{N}_1} = \frac{\dfrac{r_2}{1 - \hat{q}_2^k}}{\dfrac{r_1}{1 - \hat{q}_1^k}} \tag{4.13}$$

where the $\hat{N}_i$ are the maximum likelihood estimates (MLE) of animal abundance for the respective populations. The value of (4.13) is calculated from estimates of capture probabilities $p_1 (= 1 - q_1)$ and $p_2 (= 1 - q_2)$ estimated independently for each population by iteratively solving the equation

$$\frac{t_{2i}}{r_i} = \frac{\hat{q}_i}{1 - \hat{q}_i} - \frac{k\hat{q}_i^k}{1 - \hat{q}_i^k}$$

where

$$t_{2i} = \sum_{j=2}^{k} (j - 1)n_{ij}$$

and n_{ij} is the number of animals captured during the jth $(j = 1, \ldots, k)$ sampling period in the ith $(i = 1, 2)$ population. The estimated variance of (4.13) is again of the form of Eq. (4.12). As with the Lincoln Index and multiple-mark–recapture models, when capture rates are homogeneous between populations, proportional abundance can be estimated by (4.7) but with the variance estimate

$$\text{Var}(\hat{K}_2 \mid K) = \frac{\hat{q}^k}{r_1}(\hat{K}_2 + \hat{K}_2^2) \tag{4.14}$$

where $\hat{p} = 1 - \hat{q}$ is the maximum likelihood estimator (MLE) for the daily capture probabilities computed after pooling the capture data of the two populations (Skalski et al. 1984). As noted earlier, proportional abundance estimator (4.11) and its variance (4.12) also pertain to the generalized removal model, i.e., model M_{bh} (Otis et al. 1978; Skalski and Robson 1979, 1982), which has applications when capture probabilities vary between individuals and are affected by the trap response of the animals.

The construction of $(1 - \alpha)100\%$ confidence interval estimates of K for removal sampling can be accomplished by solving for K in the expression

$$Z_{1-(\alpha/2)} = \frac{\dfrac{K}{\hat{N}_2} - \dfrac{1}{\hat{N}_1}}{\sqrt{\dfrac{K^2 \, \text{Var}(\hat{N}_2 \mid N_2)}{\hat{N}_2^4} + \dfrac{\text{Var}(\hat{N}_1 \mid N_1)}{\hat{N}_1^4}}} \tag{4.15}$$

when (4.13) is the most appropriate estimator of K. Similarly, when K can be estimated by (4.7), an associated confidence interval can be constructed by solving for K in

$$Z_{1-(\alpha/2)} = \frac{\dfrac{K}{r_2} - \dfrac{1}{r_1}}{\sqrt{\dfrac{K^2 \, \hat{\text{Var}}(r_2 \mid N_2)}{r_2^4} + \dfrac{\hat{\text{Var}}(r_1 \mid N_1)}{r_1^4}}} \qquad (4.16)$$

where $\hat{\text{Var}}(r_i \mid N_i) = r_i \hat{q}^k$ with $\hat{p} = 1 - \hat{q}$ the pooled estimate across populations. To compute a one-tailed $(1 - \alpha)100\%$ confidence interval estimate of K using (4.15) or (4.16), a two-tailed interval at $(1 - 2\alpha)100\%$ can be calculated and the irrelevant tail estimate ignored.

EXAMPLE Analysis of a comparative census for removal sampling will be illustrated with a contrived data set (Table 4.2). The data represent captures during three days of removal trapping of the Great Basin pocket mouse (*Perognathus parvus*) on two 2.5-ha study plots in sagebrush–bunchgrass and cheatgrass environments of southeastern Washington state. A test of homogeneity indicates that capture probabilities were significantly different between populations of the two sites [test statistic $Q = 2.984$; $P(\chi_1^2 \geq 2.984) = 0.084$; see Skalski et al. (1984) for calculations]. Maximum likelihood estimates of abundance based on removal data result in population estimates of 246.20 [$\hat{\text{Var}}(\hat{N} \mid N) = 119.91$] and 172.87 [$\hat{\text{Var}}(\hat{N} \mid N) = 459.43$] animals for the sagebrush and cheatgrass environments, respectively. The population estimates consequently provide a proportional abundance estimate of

$$\hat{K} = \frac{246.20}{172.87} \approx 1.42$$

with a variance estimate of $\hat{\text{Var}}(\hat{K} \mid K) = 0.0352$. A 90% confidence interval for K can be constructed from (4.15) where

$$1.645 = \frac{\dfrac{K}{246.20} - \dfrac{1}{172.87}}{\sqrt{\dfrac{119.91 K^2}{(246.20)^4} + \dfrac{459.43}{(172.87)^4}}},$$

which simplifies to a quadratic equation in K,

$$1.6409 K^2 - 4.6990 K + 3.2070 = 0$$

Table 4.2

Summary of Capture Data of *Perognathus parvus* during 3 Days of Removal Trapping on Two 2.5-ha Study Plots

| | Number of captures | | |
Sampling period	Sagebrush	Cheatgrass	Totals
1	$n_{11} = 123$	$n_{21} = 65$	$n_{\cdot 1} = 188$
2	$n_{12} = 66$	$n_{22} = 37$	$n_{\cdot 2} = 103$
3	$n_{13} = 28$	$n_{23} = 27$	$n_{\cdot 3} = 55$
Totals	$r_1 = 217$	$r_2 = 129$	$r_{\cdot} = 346$
	$t_{21} = 122$	$t_{22} = 91$	$t_{2\cdot} = 213$

whose solutions are the endpoints to the confidence interval. The resulting confidence interval is $C.I.(1.12 \leq K \leq 1.74) = 0.90$.

This example introduced an important inferential issue. Although abundance on the sagebrush plot was substantially greater than in the cheatgrass environment $[CI(1.12 \leq K \leq 1.74) = 0.90]$, it cannot be inferred that the difference is the result of habitat alone. In order to demonstrate habitat effects, a replicated investigation is needed to compare mean abundance relative to the variance among plots treated alike.

Hypothesis Testing

Composite hypotheses of the form

$$H_0: K \geq 1 \qquad (4.17)$$

against

$$H_a: K < 1 \qquad (\text{i.e.,} \quad N_2 < N_1)$$

may be tested by using $(1 - \alpha)100\%$ one-tailed confidence intervals for proportional abundance. For instance, the null hypothesis (4.17) would be rejected at a significance level α if the numerical value 1 lies outside the bounds of a $(1 - \alpha)100\%$ upper confidence interval estimate [i.e., $P(K \leq K_U) = 1 - \alpha$] of K. Similarly, the null hypothesis

$$H_0: K \leq 1 \qquad (4.18)$$

against the alternative

$$H_a: K > 1 \qquad (\text{i.e.,} \quad N_2 > N_1)$$

would be rejected at the significance level α if the value 1 lies outside the bounds of a $(1 - \alpha)100\%$ lower-tailed confidence interval estimate of K [i.e., $P(K \geq K_L) = 1 - \alpha$].

Here again, a warning concerning interpretation of one-tailed tests and interval estimates needs to be made. Although two-tailed interval estimates can be expected to have error rates near nominal levels, error rates for one-tailed tests and interval estimates are not nearly as accurate. Although algorithms (4.9), (4.15), and (4.16) were purposefully chosen for symmetry of the error rates (Skalski et al. 1983a, 1984) in the two tails, they are not perfect. As such, an interpretation of one-tailed tests at or near the critical value should be tempered by an understanding of the behavior of the normal approximations.

EXAMPLE For the capture data in Table 4.2, a 95% lower-tailed confidence interval estimate of K can be read from the previous two-tailed interval estimate to be $CI(K \geq 1.12) = 0.95$. Because the value 1 falls below the limits of the one-tailed interval estimate, the null hypothesis (4.18) is rejected at $\alpha = 0.05$ for the comparative census. The conclusion is that pocket mouse abundance is greater on the sagebrush plot. The relative difference in abundance at the sagebrush plot is best estimated to be 1.42 times that on the cheatgrass study plot.

Specifying the Performance of Comparative Censuses

Skalski et al. (1983a) present two different design criteria useful in implementing comparative censuses. The first of these criteria is the precision of a proportional abundance estimate defined as

$$P\left(\left|\ln\left(\frac{\hat{K}}{K}\right)\right| < \epsilon\right) \geq 1 - \alpha. \tag{4.19}$$

In this form, precision of a comparative census is invariant to whether K or $1/K$ is estimated. An interesting alternative to this concept of precision is the ordinating power of an interval estimate of K. This criterion of ordinating power will be the focus in designing a comparative census.

The ordinating power is defined as the probability of properly ranking the two populations of a comparative census according to their abundance (e.g., $N_1 > N_2$ or $N_1 < N_2$). This criterion follows from the perspective that because the two populations can be assumed to have different abundance levels *a priori,* a successful comparative census should have a high probability of properly ranking the two populations. When $K = 1$ is within the bounds of a confidence interval estimate for K, one would normally conclude that no significant difference in animal numbers was detected. However, given our prior knowledge that $N_1 \neq N_2$, one would instead conclude that the comparative census simply failed to have sufficient precision to detect the difference in abundance that existed. By this definition, ordinating power of an interval estimate can be expressed as

$$
\begin{aligned}
\text{Power} &= P[(1 - \alpha)100\% \text{ confidence interval includes the true value } K_0 \\
&\quad \text{while excluding } K' = 1] \\
&= (1 - \alpha) - P(\text{interval estimate includes true value } K_0 \text{ and} \\
&\quad \text{includes } K' = 1)
\end{aligned} \tag{4.20}
$$

As defined, ordinating power of an interval estimate cannot exceed the confidence level $(1 - \alpha)$.

This concept for the power of an interval estimate can be extended to a more general definition: the probability of including the true parameter value K_0 while excluding any other specified value $K' \neq K_0$. An important special case under this general definition is to allow $K' = A_2/A_1$, where A_1 and A_2 are the real sizes of the two populations. With $K' = A_2/A_1$, power calculations can be used to design a study with a high probability of detecting a difference in the animal densities among the populations being compared. This approach is especially useful when study areas in a comparative census are of different sizes. The ordinating power of an interval estimate of proportional abundance can be written as

$$\text{Power} = (1 - \alpha) - P\left(Z_{\alpha/2} \leq Z \leq Z_{1-(\alpha/2)} - \frac{|\ln K_0 - \ln K'|}{\sqrt{\text{Var}(\ln \hat{K} \mid K)}}\right). \quad (4.21)$$

Power calculation (4.21) is appropriate with any proportional abundance estimator where $\ln \hat{K}$ can be assumed normally distributed. The variance expression $\text{Var}(\ln \hat{K} \mid K)$ can be approximated in (4.21) by the delta method as $\text{Var}(\hat{K} \mid K)/K^2$. Skalski et al. (1983a) illustrate the use of power calculation (4.21) where proportional abundance is estimated by Lincoln Indices. In the case of removal sampling, Skalski et al. (1984) illustrate applications of the ordinating power of interval estimates of K.

Design of Field Studies

Calculations of sampling effort similar to those used in abundance estimation can be performed for comparative censuses. For given values of population size N_1 and N_2, the necessary capture probability to achieve a desired level of ordinating power can be calculated. As in abundance studies, the process of sample size calculations and the subsequent field study can be viewed as confirming or refining the preliminary estimate of proportional abundance used in the design calculations.

Lincoln Index
When populations to be compared are surveyed using independent Lincoln Indices, the necessary per-period capture probability (p) is

$$p = \frac{1}{\left[1 + \dfrac{|\ln K_0 - \ln K'|}{(Z_{1-(\alpha/2)} - Z_{1-(\alpha/2)-P_W})\sqrt{\dfrac{1}{N_1} + \dfrac{1}{N_2}}}\right]}, \quad (4.22)$$

where P_W is the desired power of ordination $(P_W \leq 1 - \alpha)$, and where $\hat{K}$ is based on estimates of absolute abundance $(\hat{N})$. The value of p computed in (4.22) is calculated as if capture probabilities were heterogeneous between populations but using a common best guess (p). When K can be estimated using indices of relative abundance (i.e., $\hat{K} = r_2/r_1$), the per-period capture probability (p) is found to be

$$p = 1 - \frac{1}{\sqrt{\left[1 + \dfrac{(Z_{1-(\alpha/2)} - Z_{1-(\alpha/2)-P_W})^2 \left(\dfrac{1}{N_1} + \dfrac{1}{N_2}\right)}{(\ln K' - \ln K_0)^2}\right]}}. \quad (4.23)$$

EXAMPLE To illustrate these calculations, suppose that there are two populations of gray squirrels (*Sciurus carolinensis*) whose abundance one wishes to compare. One of these populations is in a favorable oak-hickory forest and is believed to comprise approximately 150 animals. The second population is an equal area of presumably less favorable habitat and expected to have only half the number of animals. On the basis of Lincoln Indices, a comparative census is to be designed so that the resulting 95% confidence interval estimate of proportional abundance (K) will have a 90% chance of indicating the proper ranking of the populations (i.e., $\alpha = 0.05$, $P_W = 0.90$ when $K_0 = 150/75$, $K' = 1$). On the basis of these priors, Eq. (4.22) indicates a per-period capture probability of $p = 0.410$ is needed. If capture rates can be assumed homogeneous, however, (4.23) gives $p = 0.178$. The difference in these capture probabilities indicates the relative efficiency of index methods over abundance estimation in comparative studies.

Darroch's Constant Probability Multiple-Mark–Recapture Model

For the case of comparative censuses using multiple-mark–recapture methods, sampling effort will be calculated from the parsimonious Darroch (1958) model, assuming $\hat{N}$ to be approximately log-normally distributed. For a prespecified power of ordination (P_W), the per-period capture probability necessary to compare abundance levels using $\hat{K}_1$, (4.11), is found by numerically solving for p in the equation

$$\frac{1}{(1-p)^k} + (k-1) - \frac{k}{(1-p)} = \frac{[\ln(K') - \ln(K_0)]^2}{\left(\frac{1}{N_1} + \frac{1}{N_2}\right)(Z_{1-(\alpha/2)} - Z_{1-(\alpha/2)-P_W})^2}. \qquad (4.24)$$

When capture probabilities between populations can be assumed homogeneous, the required per-period capture probability for a multiple-mark–recapture study with fixed α and P_W is found by using the equation

$$p = 1 - \left[1 + \frac{(Z_{1-(\alpha/2)} - Z_{1-(\alpha/2)-P_W})^2(K + K^2)}{N_1(\ln(K') - \ln(K))^2}\right]^{-(1/k)} \qquad (4.25)$$

Alternatively, in cases where p is considered fixed, Eqs. (4.24) and (4.25) can be solved in terms of k to determine the number of sampling periods required for a comparative census to have prescribed performance using multiple-mark–recapture surveys.

Zippin's Constant Probability Removal Model

When the populations in a comparative census are surveyed using removal sampling, power calculations in Skalski et al. (1984) can be used to determine the necessary per-period capture probability. Based on an anticipated use of absolute abundance estimators, the required per-period capture probability is found by numerically solving for p in the equation

$$\frac{(1 - (1-p)^k)(1-p)^k}{(1 - (1-p)^k)^2 - (pk)^2(1-p)^{k-1}} = \frac{\left(\dfrac{K' - K_0}{K_0}\right)^2}{(Z_{1-(\alpha/2)} - Z_{1-(\alpha/2)-P_W})^2\left(\dfrac{1}{N_1} + \dfrac{1}{N_2}\right)}$$

$$(4.26)$$

where k is the number of trapping periods. Alternatively, when K can be estimated using catch indices, the required per-period capture probability for specified α and P_W in removal sampling is

$$p = 1 - \cfrac{1}{\sqrt{1 + \cfrac{(Z_{1-(\alpha/2)} - Z_{1(\alpha/2)-P_W})^2 \left(\dfrac{1}{N_1} + \dfrac{1}{N_2}\right)}{\left(\dfrac{K'}{K_0} - 1\right)^2}}}, \tag{4.27}$$

which is of closed form.

EXAMPLE A two-sample removal technique is generally not recommended for a field study, but serves a purpose in comparison with a Lincoln Index. For the previous example of a comparative census of gray squirrel populations, the required per-period capture probabilities for a two-sample removal study are, from (4.26), $p = 0.596$ and, from (4.27), $p = 0.279$. Results again illustrate the efficiency of population comparisons based on catch indices when capture probabilities can be assumed homogeneous. In addition, these capture levels are above comparable values for the single-mark–recapture method presented earlier.

Design Optimization

Optimization of a comparative census can have various objectives. We have chosen as a goal, maximization of the ordinating power of an interval estimate of proportional abundance. Despite the difference between this goal and that of optimizing an abundance survey, the approaches to design optimization are surprisingly similar. Among the similarities is the form of their respective cost functions.

At a first glance, the presence of two study plots and the addition of interplot travel costs suggest a difference between the cost functions for comparative censuses and abundance estimation. However, because the number of study sites is prespecified, the between-site travel costs are fixed and can be subtracted from the total variable costs of doing research, C_0. Furthermore, within-plot sampling costs should simply increase by a factor of 2 without any appreciable change in the nature of the expenses. Consequently, a cost function of the form used in abundance studies (3.46) is also appropriate for optimizing a comparative census. The coefficients of the cost function naturally will change to reflect the greater costs of conducting a comparative census.

Maximizing the power (4.21) is equivalent to minimizing the $CV(\hat{K})$. If the two populations of the comparative census are prespecified and of fixed sizes N_1 and N_2, the optimization approach of Robson and Regier (1964) for the Lincoln Index should be used. Alternatively, if the two sites are randomly selected from separate strata, unconditional variances should be used in design optimization.

The $CV(\hat{K})$, incorporating both spatial variability and sampling error components associated with Lincoln Index surveys, can be expressed as

$$CV(\hat{K}) \approx \sqrt{\frac{\sigma_{N_1}^2}{\mu_1^2} + \frac{\sigma_{N_2}^2}{\mu_2^2} + \left(\frac{1}{\mu_1} + \frac{1}{\mu_2}\right)\left(\frac{1}{p} - 1\right)^2}. \qquad (4.28)$$

Then incorporating the relationships

$$\sigma_{N_i}^2 = \mu_i \eta + \mu_i^2 \omega,$$

$\mu_i = D_i A$, and (3.47), the necessary trapping effort for fixed $CV(\hat{K})$ and plot size A is

$$f = \frac{A}{c} \ln\left[1 + \frac{1}{\sqrt{\dfrac{CV(\hat{K})^2 - 2\omega}{\left(\dfrac{1}{D_1} + \dfrac{1}{D_2}\right)\dfrac{1}{A}} - \eta}}\right]. \qquad (4.29)$$

As in the optimization of single-plot studies, a graphical approach to simultaneously solving the cost function (3.49) and (4.29) can be used to find the optimal combination of f and A for fixed cost or precision.

Similarly, a Taylor series approximation to (4.29) of the form

$$f \approx \frac{A}{c\sqrt{\dfrac{CV(\hat{K})^2 - 2\omega}{\left(\dfrac{1}{D_1} + \dfrac{1}{D_2}\right)\dfrac{1}{A}} - \eta}} \qquad (4.30)$$

can be used to investigate the nature of an optimal comparative census. As with (3.52), the optimal plot size for fixed $CV(\hat{K})$ is found by noting that A_{opt} will occur at or to the left of the value that minimizes (4.30). This value for fixed $CV(\hat{K})$ is

$$A_{\text{opt}} \approx \frac{2\eta\left(\dfrac{1}{D_1} + \dfrac{1}{D_2}\right)}{CV(\hat{K})^2 - 2\omega} \qquad (4.31)$$

with a resulting trap effort per plot of

$$f_{\text{opt}} \approx \frac{A_{\text{opt}}}{c\sqrt{\eta}} \qquad (4.32)$$

and a trap density of

$$\frac{f_{opt}}{A_{opt}} = \frac{1}{c\sqrt{\eta}}.$$ (4.33)

When animal abundance is Poisson-distributed ($\eta = 1$, $\omega = 0$), Eqs. (4.31) and (4.32) reveal an important relationship between the optimal designs for an abundance survey and that of a comparative census. For Poisson-distributed abundance and the null relationship of $D_1 = D_2$, the optimal plot size for a comparative census is

$$A_{opt} = \frac{4}{D \cdot CV(\hat{K})^2}$$ (4.34)

with a corresponding trapping effort of

$$f_{opt} = \frac{4}{cD \cdot CV(\hat{K})^2}$$

and a trap density of $f_{opt}/A_{opt} = (1/c)$. Consequently, for the same level of precision (i.e., same CV), the optimal plot size in a comparative census (4.34) is twice the size needed for an abundance study (3.55) with the same trap density $(1/c)$. These results underscore the importance of not simply viewing comparative studies as a series of independent surveys of animal abundance. This optimal design for a comparative census also applies when maximizing the precision of a proportional abundance estimate (4.7).

Recommendations

The objective of conducting a comparative census is distinct from that of a population survey. Both the design and analysis of a comparative census have unique characteristics that distinguish the study from simply a series of independent population surveys. The performance of a comparative census can be substantially improved by focusing the design and analysis on this comparative role of the investigation. The inferences from a comparative study can be further improved by incorporating the following recommendations:

1. The population surveys in a comparative census should be conducted with equal sampling efforts, and simultaneously to enhance the prospects of interpopulation homogeneity.

2. Test of interpopulation homogeneity should be conducted as a first step in the analysis of comparative censuses.

3. Catch indices (i.e., r values) should be used to estimate proportional abundance if capture probabilities are shown to be homogeneous.

4. Mark–recapture surveys in a comparative census should be designed and conducted to facilitate *post hoc* model selection during subsequent data analysis.

5. Selection of appropriate survey model(s) for the analysis of comparative census data should be based on goodness-of-fit tests and the use of model selection procedures as found in Otis et al. (1978) and Cormack (1985).

6. Model selection should be conducted independently for the two surveys of a comparative census.

7. An objective function should be specified in the design of a comparative census, using either sampling precision or the ordinating power of the census.

8. In designing a comparative census, an investigator should realize that the optimal design is not simply the doubling of the effort used in a single population survey.

The above recommendations should help to improve the precision of proportional abundance estimates ($\hat{K}$). In Chapters 5 and 6, paired-plot designs will exploit these estimates of proportional abundance when assessing effects on mobile species. It will be seen that the performance of paired-plot designs will depend on both the sampling errors associated with the $\hat{K}_i$'s and the stochastic variation of wild populations through time and across the landscape.

5

Manipulative Experiments

In this chapter, statistical procedures are developed to test for treatment effects on abundance of wild populations based on results of mark–recapture surveys. These will be true experiments in the statistical sense, possessing both properties of randomization and replication of treatment conditions at the population level. Treatment differences are then assessed relative to the magnitude of variance among populations treated alike. The experiments may be concerned with evaluation of habitat and population management practices or with assessing adverse effects of habitat disturbance and environmental pollutants on animal abundance. Other experiments may be designed to test basic population processes or principles such as compensation, threshold of security, interspecies competition, and inversity.

Papers by Hurlbert (1984) and Romesburg (1981) exhort the importance of controlled or manipulated field experiments in wildlife science and ecological research. Romesburg (1981) believes that "part of wildlife science's knowledge bank has become grossly unreliable owing to the misuse of scientific methods, and major retracting is inevitable." One of the reasons, according to Romesburg, is that, "research hypotheses either are forgotten, or they gain credence and the status of laws through rhetoric, taste, authority, and verbal repetition." For some species and animal populations, manipulative experiments are neither appropriate nor feasible, and other means of assessing population effects are necessary. The point Hurlbert (1984) and Romesburg (1981) make clear is that even among species that lend themselves to experimental manipulation and studies with the stated objective of testing treatment hypotheses, manipulative experiments often are either poorly implemented or not used at all.

Current Approaches to Field Experiments

To date, little attention has been given to development of methods for analysis of capture data in a hypothesis testing framework. This paucity of quantitative guidance has resulted in many field studies not achieving their full
potential. To appreciate the necessity for better approaches to the design
and analysis of multiplot, mark–recapture studies, a review of current
methods is necessary.

Small mammals, more than any other group of species, are subjected to
experimental manipulation in wildlife science. These studies provide an opportunity to evaluate the state of the art of field experimentation in wildlife
ecology. Yet, even among these species that are most amenable to experimental manipulation, basic design flaws often are present (Table 5.1). A
survey of 30 manipulative experiments on small-mammal abundance (Table
5.1) indicated that fully one-half failed to provide replication at the population level for all treatments in the study. An even more startling observation
is that only 1 in 15 experiments used or mentioned that randomization was
employed in assigning test conditions to populations. These findings are
consistent with those of Eberhardt (1978), who found that "field experiments in ecology either have no replication, or have so few replicates as to
have very little sensitivity." Thus, it appears that despite the pretense of
testing effects on wild populations, many researchers in ecology are oblivious to the need for sound experimental principles as stated by R. A. Fisher
(1947) over four decades ago.

An important reason for using manipulative experiments is to ensure
that treatment effects are not confounded with extraneous environmental
factors. This usually requires the standardization of test conditions. In 7 of
the 30 small-mammal studies (Table 5.1), however, sizes of test plots and
trapping intensity (number of traps per unit area) were different between
treatments. These are particularly poor design configurations because effects
on abundance are assessed using catch indices that assume a constant proportionality between the index and the population level. The majority (26
of 30) of studies reported in Table 5.1 used indices of abundance to quantify
population levels. In so doing, the researchers have implicitly assumed that
either treatment effects did not change capture rates among populations
compared or that animal abundance was completely enumerated in all populations tested. Neither assumption is as universally plausible as would be
suggested by the frequency of assertions.

The assumption of complete enumeration often must be viewed as sus-
pect in field experiments, for rarely will a trapping study ensure that every
animal in a population will be captured with a probability of 1. In practice,
high capture probabilities but incomplete enumeration may bias treatment
contrasts very little. The most important reason for not using complete

Table 5.1

Survey of Manipulative Experiments on Small-Mammal Abundance with Summaries of the Treatment Design Used and Levels of Effort Employed

Reference	Treatment design	Survey method	Rand.*	Level rep.† (P/T)	Plot size (ha)	Trapping effort (traps/plot)	Species
LoBue and Darnell 1959	Pre- and postalfalfa cutting	Total catch measured	No	1	1.98	225 traps (128.6 ha^{-1})	*Microtus pennsylvanicus, Peromyscus maniculatus*
Krebs and DeLong 1965	Control and food supplement	Calendar of captures	No	1	Control 0.82, treatment 0.70	126 traps (379.7 ha^{-1}), 200 traps (287.3 ha^{-1})	*Microtus californicus, Mus musculus*
Tester 1965	Pre- and postburning	Total catch measured	No	1	8.09	120 traps (14.8 ha^{-1})	*Peromyscus maniculatus, Cleithrionomys grapperi*
Ahlgren 1966	Control, cut, cut and burned	Total catch measured	No	1–2	4.05	60 traps (14.8 ha^{-1})	*Peromyscus maniculatus, Cleithrionomys grapperi*
Barrett and Darnell 1967	Control and two levels of dimetholate	Schnabel (1938) estimates	No	1	1.07	90 traps (84.2 ha^{-1})	*Microtus ochrogaster, Mus musculus, Peromyscus maniculatus*
Fordham 1971	Control and food supplement	Total catch measured	No	3	0.6–2.0	8–12 traps (4–20 ha^{-1})	*Peromyscus maniculatus*
Beck and Vogi 1972	Control 2, 4, and 11 repeated burns in 15 years	Total catch measured	No	1	3.36	50 traps (14.9 ha^{-1})	*Peromyscus maniculatus, Cleithrionomys grapperi, Peromyscus leucopus*
Flowerdew 1972	Control and food supplement	Calendar of captures	No	1	Control 1.5, treatment 1.1	90 traps (60 ha^{-1}), 90 traps (81.8 ha^{-1})	*Apodemus sylvaticus*
Joule and Jameson 1972	Control and two levels of species removal	Total catch measured	No	3–4	1.2	54 traps (45 ha^{-1})	*Sigmodon hispidus, Reithrodontomys fulvescens, Oryzomys palustris*
Hoover 1973	Control, logged, logged and slash burned	Total catch measured	No	1	Control 2.02, treatment 4.04	100 traps (49.5 ha^{-1}), 200 traps (49.5 ha^{-1})	*Microtus oregoni, Peromyscus maniculatus*

(continues)

Table 5.1 (*continued*)

Reference	Treatment design	Survey method	Rand.*	Level rep.[†] (P/T)	Plot size (ha)	Trapping effort (traps/plot)	Species
Rosenzwig 1973	Control, cleared, and augmented vegetation	Total catch measured	No	3	0.08	24 traps (298.4 ha^{-1})	*Perognathus penicillatus, Dipodomus merriami*
Andrzejewski 1975	Control and two levels of supplemental food	Total catch measured	No	1	2.70	117–247 traps (43.3–91.5 ha^{-1})	*Clethrionomys glareolus*
Pomeroy and Barrett 1975	Control and Sevin	Calendar of captures	No	1	0.41	128 traps (316.3 ha^{-1})	*Sigmodon hispidus, Mus musculus, Peromyscus polionotus*
Schroder and Rosenzwig 1975	Control and four levels of population reduction of *Dipodomys ordii* and *Dipodomys merriami*	Lincoln index estimates	No	2	16.2	331 traps (20.4 ha^{-1})	*Dipodomys ordii, Dipodomys merriami*
Boonstra and Krebs 1977	Control and fenced	Calendar of captures	No	1	Control 0.58, treatment 0.28	150 traps (258.6 ha^{-1}), 74 traps (264.3 ha^{-1})	*Microtus townsendii*
Cameron 1977	Control and two levels of population reduction	Calendar of captures	No	2	1.6	81 traps (50.6 ha^{-1})	*Sigmodon hispidus, Reithrodontomys fulvescens*
Grant et al.1977	2 × 2 factorial design of nitrogen and water supplements	Jolly (1965) and Zippin (1956) estimates	No	2	1.0	42 traps (42 ha^{-1})	*Peromyscus maniculatus, Microtus ochrogaster, Onychomys leucogaster*
Abramsky 1978	2 × 2 factorial design of nitrogen and water supplement plus food addition treatment	Calendar of captures	No	1–2	1.0	42 traps (42 ha^{-1})	*Peromyscus maniculatus, Microtus ochrogaster, Onychomys leucogaster*
Hansen and Batzli 1978	Control and food supplement	Calendar of captures	No	1	Control 1.8, treatment 1.4	88 traps (48.9 ha^{-1}), 72 traps (51.4 ha^{-1})	*Peromyscus leucopus*
Crowner and Barrett 1979	Control and burned areas	Calendar of captures	No	1	0.4	100 traps (250 ha^{-1})	*Mus musculus, Peromyscus maniculatus, Microtus pennsylvanicus*

Reference	Treatment design	Measurement method	Randomization[*]	Level of replication[†]	Density	Trapping effort	Species
Gaines et al. 1979	Control, removal trapping, and fenced	Calendar of captures	No	2–3	0.8	100 traps (125 ha⁻¹)	*Microtus ochrogaster*
Langley and Shure 1980	Pine and abandoned fields	Schumacher-Eschmeyer (1943)	No	2	1.28	50 traps (39.1 ha⁻¹)	*Sigmodon hispidus*
Munger and Brown 1981	2 × 2 factorial design of harvest ant and large rodent removal	Total catch measured	Yes	2	0.25	49 traps (196 ha⁻¹)	*Perognathus penicillatus, Peromyscus maniculatus, Onychomys leucogaster*
O'Meara et al. 1981	Chained 1, 8, 15 years ago plus control	Total catch measured	No	3	0.72	64 traps (88.9 ha⁻¹)	*Peromyscus maniculatus, Eutamias minimus*
Tait and Krebs 1981	Control and five levels of food supplement	Calendar of captures	No	1–2	0.21, 0.47	50 traps (238.1 ha⁻¹), 100 traps (212.8 ha⁻¹)	*Microtus townsendii*
Anderson and Barrett 1982	Control, sludge-treated, and commercial fertilizer	Calendar of captures	Yes	4–6	0.10	25 traps (250 ha⁻¹)	*Microtus pennsylvanicus*
Fox 1982	Control and burned	Calendar of captures	No	1	Control 4.0, treatment 7.0	96 traps (24 ha⁻¹), 177 traps (25.3 ha⁻¹)	*Mus musculus, Pseudomys novaehollandiae, Pseudomys gracilicaudatus*
Halvorson 1982	Control and burned	Total catch measured	No	2	1.88	81 traps (43.1 ha⁻¹)	*Peromyscus maniculatus, Cleithrionomys grapperi, Eutamias ruficaudus*
Grant et al. 1982	Control and grazed	Density index based on total captures	No	4	2.70	144 traps (53.3 ha⁻¹)	*Peromyscus maniculatus, Onychomys leucogaster*
Desy and Thompson 1983	Control, food supplement, and removal trapping	Calendar of captures	No	1	0.8	56 traps (70 ha⁻¹)	*Microtus pennsylvanicus*

[*] Randomization indicated.

[†] Level of replication (plots per treatment).

emumeration in field studies is the inefficient use of sampling effort it represents. As will be seen later, the optimal level of trapping intensity in a multiplot, mark–recapture experiment may be far below the level necessary for complete enumeration of animal abundance. The additional field effort often is better utilized by increasing the level of field replication in the experiment. Nichols and Pollock (1983) provide a good discussion of the complicated interpretation of the calendar-of-captures method. The calendar-of-captures method suffers from the same inefficient use of sampling effort as complete enumeration, and for this reason should be avoided.

Any treatment that may affect demographic structure of a population or animal behavior should be suspected of possibly changing capture rates, as well as abundance levels. Differences in trappability have been reported among different age and sex classes of animals (Young et al. 1952; Huber 1962; Gliwicz 1970). Catch rates of gray squirrels (*Sciurus carlinensis*), for example, also have been shown to be affected by changes in habitat (Perry et al. 1977). Changes in habitat may be particularly important in sighting studies where the rate of detection may be influenced by changes in habitat structure and visibility. Furthermore, when replicate populations in a field experiment are surveyed over time, weather factors (Getz 1961; Mystkowska and Sidorowicz 1961; Gentry et al. 1966) and seasonal changes (Klonglan 1955; Perry et al. 1977) may affect the rates of capture in trapping studies.

To differentiate changes in capture probabilities from effects of treatments on abundance, capture data must permit tests of homogeneity. Often, the nature of a catch index does not lend itself to verification of the assumption of homogeneity unless the trapping study is concurrently designed for estimation of absolute abundance. Thus, unless tests of homogeneity have verified the appropriateness of using indices of abundance, most of the field experiments reported in Table 5.1 risked confounding treatment effects on animal abundance with possible changes in capture rates.

Unequal plot sizes further complicate the interpretation of test results because the effective trapping area about a study plot usually is not viewed as a linear function of plot size (Figure 5.1). If plot sizes must vary in an experiment, plot sizes should be randomized to the treatments or the principle of blocking used, and techniques for density estimation employed [see Otis et al. (1978)].

Considerations in Choice of Statistical Tests

The uncertain nature of animal dispersion across the landscape makes specification of exact tests impractical. Skalski (1985a, pp. 216–221) inves-

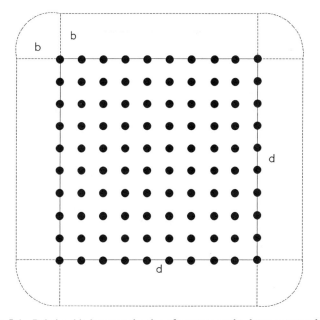

Figure 5.1 Relationship between the size of a square study plot as measured by the outside rows of traps (d^2) and the effective plot size measured as having a border strip of width b determined by the extent of movement of animals about the study plot ($d^2 + 4bd + \pi b^2$).

tigated distributional properties of a likelihood ratio test (LRT) of treatment effects based on analysis of capture data, assuming that animal abundance was Poisson-distributed. The test statistic was well behaved, being nominally chi-square-distributed with even minimal replication (two replicate plots in each of two treatments) when animal abundance was Poisson-distributed. However, the test procedure was very sensitive to departures from the assumptions of Poisson-distributed animal abundance. Computer simulations showed that the α-levels of this LRT were severely affected by binomial and negative binomial distributions. Because the processes that affect dispersion of vertebrate populations are poorly understood and dispersion patterns can be expected to change with experimental conditions and environmental factors, the use of parametric assumptions seems unwarranted.

The nonnormality and uncertain distributional properties of animal abundance may suggest the use of distribution-free statistical methods such as permutation or randomization tests, which are robust to the underlying distribution of the data (Edgington 1980). These statistical tests of effects, however, require a random assignment of test conditions to experimental units, i.e., the study sites. Consequently, use of these test procedures is restricted to the current category of manipulative experiments.

The number of replicate populations per treatment will influence the potential usefulness of randomization tests. For example, in a two-treatment completely randomized design with $l = 2$ replicate sites per treatment, there are $\binom{4}{2} = 6$ possible randomizations of the data. These six outcomes restrict a one-tailed test to a minimum α level of 0.167 or a two-tailed test to $\alpha = 0.333$. These α-levels are too coarse to provide a reasonable interpretation of field data, although they do suggest the inferential limitations of experiments with too little replication. Not until $l = 4$ replicates, can a two-treatment randomization design provide a two-tailed significance test at $\alpha \leq 0.05$. Skalski (1985a, pp. 220–233) found that normal approximations to a two-treatment randomization test (Conover 1971, pp. 357–369; Scheffé 1959, pp. 313–329) for a completely randomized design approached the nominal test distribution only after l reaches eight or more replicates per treatment. Consequently, at levels of field replication typically encountered in wildlife experiments, researchers should use the actual randomization process rather than normal approximations to randomization tests in analyzing manipulative experiments.

In the remainder of this chapter and in Chapter 6, statistical analyses will depend on the robustness of standard statistical tests, assuming normality under various transformations of the data. Computer simulations of mock mark–recapture studies (Skalski 1985a) were used to verify the distributional behavior of these test statistics under a variety of assumed distributions for animal abundance. These simulation results suggest that within the context presented, researchers can make valid inferences to treatment effects on animal abundance when mark–recapture data are analyzed by the test procedures described below. The remarkable robustness of normal theory tests permit investigators to make inferences at α-levels such as 0.01 and 0.05, when randomization tests at these same α-levels would be precluded. Ability to perform near exact α-level tests with small to moderate replication is the result of approximate normality of the treatment contrasts investigated.

In the following sections of this chapter, tests of hypotheses are presented for data collected from completely randomized and paired experimental designs with two treatments. These designs hardly will satisfy all experimental needs, but seem to most closely represent the majority of experimental approaches represented in Table 5.1. Replicate population surveys in an experiment will be illustrated using independent single-mark–recapture surveys (i.e., Lincoln Indices) at each site. Extension of the methodology to the broader class of closed-population abundance estimators is straightforward. The structure of the sampling error will change and alternative tests of homogeneity will be required depending on the survey model. However, the general principles of design and analysis of multiplot experiments will remain the same. We generally recommend a ln-transformation of the abundance estimates or catch indices prior to analysis unless auxiliary distributional information is available.

Completely Randomized Two-Treatment Experimental Design

Scenario

The experimental design consists of $l_C + l_T$ study plots, each of equal area. To these plots, the control (C) and treatment (T) conditions are randomly assigned. Sometime after application of treatments, population levels are estimated at each plot using Lincoln Index surveys to test for treatment effects on animal abundance. For the statistical inference to reflect the desired experimental inference, the $l_C + l_T$ study sites need to be sampled from habitat or locations of interest. No form of statistical analysis will make up for improper site selection at this stage of a manipulative experiment.

The null hypothesis to be tested is that there is no difference in mean abundance (μ_N) between control and treatment populations. In other words, the treatment has no effect on mean abundance of the animal, say

$$H_0: \mu_C = \mu_T \tag{5.1}$$

against the alternative hypothesis of

$$H_a: \mu_C \neq \mu_T$$

If *a priori* knowledge exists about the potential treatment effect, a test with a one-sided alternative hypothesis may be tested.

The scenario for a randomized experiment should not be equated with the case of measuring differences in mean abundance between two different habitat types (Figure 5.2) called a "planned survey" (Cox 1958, p. 2). In

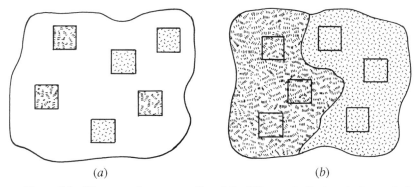

(a)   (b)

Figure 5.2 Diagrammatic representation of the difference in the layout of a manipulative experiment (a) and planned survey (b).

manipulative experiments, the treatments being compared are randomly assigned to study plots by the investigator. Treatments being tested may indeed be different habitats, but they must be randomly assigned to study plots through a chance process. In a planned survey, the investigator has no control over which experimental units (i.e., the study sites) fall into one category or another. In other words, the treatments are self-assigned in a planned survey. Valuable information can be obtained from planned surveys, but unobscured conclusions about causal effects are best obtained from manipulative experiments.

The distinction between a manipulative experiment and a planned survey has particular importance in analyses of wildlife studies. In Chapter 2, it was suggested that changes in dispersion ($\sigma_{\hat{N}}^2$) may be expected to accompany changes in mean abundance of a species. For some treatments, the principal population effect may be a direct effect on the dispersion pattern of an animal population. In either situation, unequal treatment variances may result, a condition called "heteroscedasticity." Unequal treatment variances may affect the rate the null hypothesis is rejected and produce α levels far from expected. Historically, this is known as the "Behrens–Fisher problem" (Behrens 1929; Fisher and Yates 1957).

The Behrens–Fisher problem is a nonproblem under randomization. In manipulative experiments, if treatments have no effect on animal abundance, treatment variances will have the same expected value under the null hypothesis (5.1). In planned surveys, differences in mean response can be expected between domains, and thus, the objective of the survey should be to estimate the size of the unknown difference. Consequently, the assumption of homogeneous treatment variances is unfounded and heterogeneity can be expected. Hence, from a philosphical as well as a practical standpoint, the distinction between manipulative experiments and planned surveys is important. A detailed analysis of planned surveys will not be discussed in this book [see Skalski (1985a, pp. 166–183)].

Test Statistics

The model equation for this completely randomized two-treatment design can be written as

$$\hat{N}_{ij} = \mu \tau_i \epsilon_{ij} \tag{5.2}$$

where $\hat{N}_{ij}$ = estimated animal abundance at the jth replicate plot
$(j = 1, \ldots, l_i)$ of the ith treatment $(i = C, T)$
μ = overall mean
τ_i = effect of the ith treatment $(i = C, T)$
ϵ_{ij} = multiplicative error associated with the jth plot
$(j = 1, \ldots, l_i)$ and ith treatment $(i = C, T)$

It is fortuitous that the same data transformation that linearizes the multiplicative response model (5.2) also results in the approximate normality of the Lincoln Index and the associated catch index (r). Using a logarithmic transformation, the response model becomes

$$\ln \hat{N}_{ij} = \ln \mu + \ln \tau_i + \ln \epsilon_{ij}.$$

This model expresses the belief that various treatments have a multiplicative effect on mean abundance.

The most appropriate statistic to test the null hypothesis (5.1) depends on whether the data set is balanced and whether capture probabilities are homogeneous among populations. A balanced design results in a test statistic that is robust to heterogeneity in treatment variances. Between-plot variances can be expected to change with differences in mean abundance and capture probabilities. For this reason, a balanced design is very important because only very approximate test procedures for unbalanced data exist in analyzing capture data.

Patterns of interpopulation homogeneity can be classified into three categories; these are

Case 1: Different probabilities $(p_{1k}, p_{2k}$ for $k = 1, \ldots, l_c + l_T)$ exist for all $l_C + l_T$ populations.

Case 2: Capture probabilities $(p_1$ and $p_2)$ are homogeneous among replicate populations within a treatment but differ between treatments.

Case 3: Capture probabilities $(p_1$ and $p_2)$ are homogeneous for all $(l_C + l_T)$ populations.

The proper test statistic will depend on which of the three cases best represents the experimental scenario. Chi-square tests of homogeneity using contingency tables (3.11) may be used to test the assumption of homogeneous capture probabilities and determine the proper test of the null hypothesis (5.1). For surveys based on multiple-mark–recapture methods and removal sampling, contingency tables (3.15) and (3.20) should be used, respectively.

Balanced Design with Case 1   When the capture probabilities are heterogeneous among all the populations in the experiment, estimates of absolute abundance based on the Lincoln Index must be used to test the null hypothesis. The statistic may be written as

$$d = \frac{\bar{X}_{C\cdot} - \bar{X}_{T\cdot}}{\sqrt{\dfrac{S_{\bar{X}_C}^2}{l_C} + \dfrac{S_{\bar{X}_T}^2}{l_T}}}, \tag{5.3}$$

where

$$X_{ij} = \ln \hat{N}_{ij},$$

$$\bar{X}_{i\cdot} = \sum_{j=1}^{l} \frac{\ln \hat{N}_{ij}}{l},$$

$$S_{\bar{X}_i}^2 = \frac{\displaystyle\sum_{j=1}^{l} (\ln \hat{N}_{ij})^2 - \frac{\left(\displaystyle\sum_{j=1}^{l} \ln \hat{N}_{ij}\right)^2}{l}}{l - 1},$$

and is approximately distributed as a t-statistic with ($l_C + l_T - 2 = 2l - 2$) degrees of freedom.

Balanced and Unbalanced Designs with Case 2 In this case, all replicate plots within a treatment are assumed to exhibit the same vector of capture probabilities for the Lincoln Indices. When this is true, mark–recapture data from the replicate plots may be combined to provide more precise estimates of capture probabilities. These estimates of p_1 and p_2 can be used, in turn, to more precisely estimate animal abundance at each plot. The resulting test statistic can be written as

$$Z = \frac{\hat{\mu}_C - \hat{\mu}_T}{\sqrt{\text{Vâr}(\hat{\mu}_C) + \text{Vâr}(\hat{\mu}_T)}} \tag{5.4}$$

where

$$\hat{\mu}_i = \frac{n_{1\cdot} n_{2\cdot}}{m \cdot l},$$

$$\text{Vâr}(\hat{\mu}_i) = \frac{S_{r_i}^2 n_{1\cdot}^2 n_{2\cdot}^2}{m^2 r_{\cdot}^2} + \frac{(n_{1\cdot} - m_{\cdot})(n_{2\cdot} - m_{\cdot})(r_{\cdot} - m_{\cdot}) n_{1\cdot} n_{2\cdot}}{m_{\cdot}^3 r_{\cdot}^2}$$

and where the

$$S_{r_i}^2 = \frac{\displaystyle\sum_{j=1}^{l} r_{ij}^2 - \frac{\left(\displaystyle\sum_{j=1}^{l} r_{ij}\right)^2}{l}}{l - 1}$$

are computed separately for control and treatment plots ($i = C, T$). Test statistic (5.4) has approximately a standard normal distribution under the null hypothesis (5.1). It is important to note that (5.4) is based on the assumption of an additive response model. In the case of multiplicative effects, the relationship $\text{Vâr}(\ln(\hat{\mu})) \approx \text{Vâr}(\hat{\mu})/\hat{\mu}^2$ may be used to construct a test statistic consistent with (5.2). The resulting test statistic based on a multiplicative response model is

$$Z = \frac{\ln(\hat{\mu}_C) - \ln(\hat{\mu}_T)}{\sqrt{\dfrac{\hat{V}ar(\hat{\mu}_C)}{\hat{\mu}_C^2} + \dfrac{\hat{V}ar(\hat{\mu}_T)}{\hat{\mu}_T^2}}} \tag{5.5}$$

which is also approximately distributed as a standard normal variate. Simulation results (Skalski 1985a, pp. 161–165) suggest a need for $\mu_C = \mu_T > 2$ and $p > 0.2$ so that statistic (5.4) is normally distributed. When these minimal conditions for the valid use of test statistics (5.4) and (5.5) are not met, the capture data should be analyzed using (5.3). The advantage in the use of test statistic (5.4) or (5.5) over test statistic (5.3) is the greater power to reject the null hypothesis (5.1) when treatment effects indeed exist.

Balanced Design with Case 3   In this case, all $2l$ populations exhibit homogeneous capture probabilities and consequently, population comparisons can be based on the catch indices, $r_{ij}(i = 1, 2, j = 1, \ldots, l)$. The test statistic is the same as (5.3) except the r_{ij} are used instead of the $\hat{N}_{ij}$. The statistic can be written as

$$d = \frac{\bar{Y}_{C\cdot} - \bar{Y}_{T\cdot}}{\sqrt{\dfrac{S_{\bar{Y}_C}^2}{l_C} + \dfrac{S_{\bar{Y}_T}^2}{l_T}}} \tag{5.6}$$

where

$$\bar{Y}_{i\cdot} = \frac{\displaystyle\sum_{j=1}^{l} \ln r_{ij}}{l}$$

$$S_{\bar{Y}_i}^2 = \frac{\displaystyle\sum_{j=1}^{l} (\ln r_{ij})^2 - \dfrac{\left(\displaystyle\sum_{j=1}^{l} \ln r_{ij}\right)^2}{l}}{l - 1}$$

and is approximately distributed as a t-statistic with $2l - 2$ degrees of freedom. In both case 1 and case 3, untransformed capture data result in test statistics that are approximately t-distributed. However, the power of the statistics with ln-transformed data can be appreciably greater under H_a when treatments have a multiplicative effect on population levels.

Unbalanced Designs with Cases 1 and 3   Heterogeneity in sampling variances (result of unequal capture probabilities or unequal mean abundance levels) or plot-to-plot variances (as in the case of planned surveys) becomes critical when there are unequal numbers of replicate control plots (l_C) and treatment plots (l_T). This condition should always be avoided if possible. However, uncontrollable circumstances may result in unbalanced designs. For this reason, procedures for unbalanced designs are considered.

For an unbalanced design, the interpretation of the d-statistics [(5.3), (5.6)] using the adjustment of Cochran (1964) was found to perform adequately, whereas the approach suggested by Welch (1938) tended to be liberal (observed $\alpha >$ nominal levels). The difference between the methods of Cochran (1964) and Welch (1938) is in the calculation of an adjusted α-level. Only the approach suggested by Cochran will be considered here. For d-statistics (5.3) and (5.6), the adjusted critical value is calculated as

$$
t_\alpha' = \frac{\dfrac{s_C^2}{l_C} \cdot t_{\alpha, l_C - 1} + \dfrac{s_T^2}{l_T} \cdot t_{\alpha, l_T - 1}}{\dfrac{s_C^2}{l_C} + \dfrac{s_T^2}{l_T}} \tag{5.7}
$$

where $t_{\alpha, l_C - 1}$ is the critical value associated with a t-statistic with $l_c - 1$ degrees of freedom [i.e., $P(t \geq |t_{l_C - 1}|) = \alpha$] and $t_{\alpha, l_T - 1}$ is the critical value associated with a t-statistic with $l_T - 1$ degrees of freedom [i.e., $P(t \geq |t_{l_T - 1}|) = \alpha$]. The calculated values of the d-statistics are then compared with t_α' instead of the usual tabular values found in a table for a t-distribution. It is important to note that when the treatment with the greatest variance (e.g., larger σ_N^2 or smaller p) also has the fewest replicates, no procedure for the analysis of an unbalanced design performs adequately (Skalski 1985a, pp. 166–172).

Robustness of the Test Statistics   The d-statistics (5.3) and (5.6) assume that data used in the analyses are normally distributed. Computer simulations indicate (Skalski 1985a) the test statistics are approximately t-distributed despite population abundance (N) that is binomial, Poisson, or negative-binomially distributed. This robustness is exhibited in studies with as few as two replicate sites per treatment (two degrees of freedom) in the case of balanced designs. The results are comforting because the pattern of animal dispersion often will be unknown and its characterization, too strenuous.

There are two forms of heterogeneity in capture probabilities that investigators must consider in analysis of multiplot mark–recapture experiments (Figure 5.3). The only type of heterogeneity thus far discussed has been interpopulation heterogeneity, which determines the form of test statistic [(5.3), (5.4), or (5.6)] that must be used to test the null hypothesis (5.1). However, the Lincoln Index assumes that all individuals within a population have the same probability of capture, and it can be quite sensitive to violations of this assumption. Fortunately, in manipulative experiments, the primary objective is a test of effects and not abundance estimation.

In manipulative experiments, heterogeneity in capture probabilities among the individual animals within a population (i.e., intrapopulation

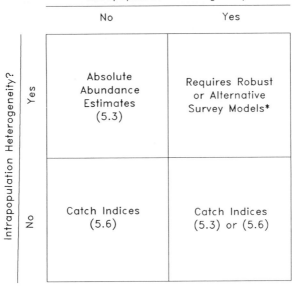

Intrapopulation Heterogeneity?

	No	Yes
Intrapopulation Heterogeneity? — Yes	Absolute Abundance Estimates (5.3)	Requires Robust or Alternative Survey Models*
Intrapopulation Heterogeneity? — No	Catch Indices (5.6)	Catch Indices (5.3) or (5.6)

* Test procedures not yet available in all cases

Figure 5.3   Valid test procedures in the presence of intra- and/or interpopulation heterogeneity in the probabilities of capture.

heterogeneity) is of concern only when it coexists with interpopulation heterogeneity (Figure 5.3). As long as the distribution of individual capture probabilities is the same among the populations being compared in a manipulative experiment, any of the test statistics [(5.3), (5.4), (5.5), or (5.6)] will provide valid tests of effects. In the presence of interpopulation homogeneity, tests of hypotheses based on use of catch indices [(5.4), (5.6)] have greater power to detect treatment effects than do their counterparts based on absolute abundance (5.3). Simulation studies in which individual capture probabilities were assumed to be beta-distributed (Figure 5.4) showed no effect on observed α-levels (Table 5.2) for d-test statistic (5.3) or (5.6). Indeed, simulation studies indicate no effect on the distribution of test statistics (5.3) and (5.6) under any of the eight mark–recapture models proposed by Otis et al. (1978) as long as the expected pattern of intrapopulation heterogeneity was the same between populations.

The multiple-mark–recapture models of Otis et al. (1978) or Cormack (1985) are generally recommended to provide valid abundance estimates before treatments are compared. Because not all forms of intrapopulation het-

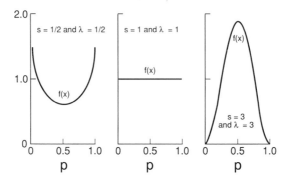

Figure 5.4 Beta probability densities used in simulation studies to investigate the robustness of tests of treatment effects to intrapopulation heterogeneity in the presence of interpopulation homogeneity.

erogeneity can currently be handled [i.e., models M_{tb}, M_{th} and M_{tbh} in Otis et al. (1978)], much more work will be necessary before valid tests of effects can be performed under the case of dual heterogeneity. In the interim, field experiments that maximize prospects of interpopulation homogeneity are essential. And in this regard, it is fortunate that the chi-square tests of interpopulation homogeneity presented in Chapter 3 remain valid under all eight of the mark–recapture models of Otis et al. (1978). Generalizations of test statistics (5.3) and (5.6) to the broad class of survey models is straightforward. Extension of test statistics (5.4 and 5.5) to other survey models must be done on a case-by-case basis using the total variance law (Appendix 1).

Example Analyses

Effects of Control Burning   The first example represents capture data from Lincoln index surveys of the deer mouse (*Peromyscus maniculatus*) on four replicate 2.5-ha study plots in old field control sites and on study plots with controlled burns 2 weeks prior to the start of the trapping program (Table 5.3). The captures represent the results from 6 consecutive days of trapping per period. Populations on all eight study plots were trapped simultaneously to enhance prospects that capture probabilities would be homogeneous between sites. The eight sites represent a random selection of 16 potential study areas, chosen for their similar vegetation and history of agricultural use. Among the eight sites, four chosen areas were randomly selected for control burning. The purpose of the study is to test the null hypothesis

$$H_0: \mu_C \leq \mu_T$$

Table 5.2

Observed α-Levels for Two-Tailed Tests ($\alpha = 0.05$ Lower One-Tailed Tests) Using the d-Statistics (5.3) and (5.6), Assuming They Were t-Distributed with $2l - 2$ Degrees of Freedom.*

Control treatment		Replications	Beta		Case 1—α				Case 3—α			
μ_N	σ_N^2	l	s	λ	LT(0.05)†	0.05	0.01	0.10	LT(0.05)†	0.05	0.01	0.10
50	100	2	1	1	0.0405	0.0390	0.0105	0.0890	0.0395	0.0350	0.0035	0.0790
		4			0.0505	0.0575	0.0095	0.1075	0.0465	0.0560	0.0095	0.1065
		8			0.0510	0.0570	0.0065	0.1055	0.0545	0.0555	0.0120	0.1050
50	100	2	3	3	0.0405	0.0510	0.0065	0.0905	0.0385	0.0375	0.0030	0.0845
		4			0.0430	0.0535	0.0100	0.0970	0.0425	0.0530	0.0110	0.1015
		8			0.0580	0.0585	0.0085	0.1140	0.0580	0.0665	0.0090	0.1170
50	100	2	$\frac{1}{2}$	$\frac{1}{2}$	0.0490	0.0470	0.0080	0.0960	0.0435	0.0395	0.0045	0.0955
		4			0.0475	0.0530	0.0115	0.0970	0.0485	0.0495	0.0095	0.0990
		8			0.0505	0.0540	0.0100	0.1030	0.0580	0.0520	0.0095	0.1040
200	1000	2	1	1	0.0515	0.0485	0.0085	0.1030	0.0485	0.0415	0.0070	0.0995
		8			0.0395	0.0415	0.0065	0.0885	0.0425	0.0485	0.0070	0.0925
200	1000	2	3	3	0.0515	0.0520	0.0105	0.1075	0.0525	0.0510	0.0075	0.1035
		8			0.0450	0.0475	0.0080	0.0945	0.0430	0.0480	0.0065	0.0990

*Animal abundance was distributed as negative-binomial and capture probabilities beta-distributed. Results based on 2000 computer simulations per scenario.

† Lower one-tailed test at $\alpha = 0.05$.

Table 5.3

Capture Data, Abundance Estimates (3.5), and Catch Indices from Lincoln Index Surveys
of Deer Mice (*Peromyscus maniculatus*) on Four-Replicate, 2.5-ha Plots under Control
and Treated (Burned) Conditions

	Control sites					Treatment sites			
	n_1	n_2	m	r	$\hat{N}$	n_1	n_2	m	r	$\hat{N}$
	49	82	26	105	152.704	19	28	9	38	57.000
	40	68	23	85	116.875	16	34	10	40	53.091
	46	66	19	93	156.450	22	25	10	37	53.364
	35	51	18	68	97.526	22	48	11	59	92.917
Totals	170	267	86	351		79	135	40	174	

against the one-tailed alternative

$$H_a: \mu_C > \mu_T.$$

For the *Peromyscus* data (Table 5.3), tests of homogeneity, Eq. (3.11),
indicate that capture probabilities were homogeneous among the replicate
control plots ($\chi_6^2 = 2.975$, $\alpha = 0.812$), among the replicate treatment sites
($\chi_6^2 = 6.507$, $\alpha = 0.369$), and between control and treatment populations
($\chi_2^2 = 0.429$, $\alpha = 0.807$). As a consequence, a test of the effect of a
2-week-old burn on deer mice abundance can be performed using (5.6),
where

$$d = \frac{4.46128 - 3.75373}{\sqrt{\dfrac{0.033670}{4} + \dfrac{0.047647}{4}}} = 4.9688$$

indicating a significant [$P(t_6 > 4.9688) = 0.001$] reduction in abundance
at treatment sites.

Effects of Chaining   The second example illustrates a case where the treatment is readily expected to have a beneficial effect on population abundance. Data from Lincoln Index surveys (Table 5.4) represent the possible
outcome of an experiment to test the effect of chaining practices on subsequent small mammal abundance. The study is to test the null hypothesis

$$H_0: \mu_C \geq \mu_T$$

against

$$H_a: \mu_C < \mu_T.$$

Table 5.4

Capture Data, Abundance Estimates (3.5), and Catch Indices from Lincoln Index Surveys
of Least Chipmunk (*Eutamias minimus*) on Four-Replicate, 6.0-ha Plots under Control
and Treated (Chained) Conditions

	Control sites				Treatment sites				
	n_1	n_2	m	r	$\hat{N}$	n_1	n_2	m	r	$\hat{N}$
	8	26	2	32	80.000	18	65	7	76	155.750
	32	48	23	57	66.375	49	73	34	88	104.714
	32	47	15	64	98.000	36	63	21	78	106.636
	30	29	11	48	76.500	31	56	16	71	106.294
Totals	102	150	51	201		134	257	78	313	

Eight 6.0-ha study sites were selected, and a random sample of four sites
was identified to receive the test condition under investigation. Capture data
represent the cumulative results of 6 days of trapping effort at each site: 3
days for initial capture and marking followed by 3 days for recapture. Unlike the previous example, the eight sites were surveyed sequentially with
the order of the surveys randomly determined.

Tests of homogeneity indicate heterogeneous capture probabilities
among replicate control sites ($\chi^2_6 = 22.330$, $\alpha = 0.001$), replicate treatment sites ($\chi^2_6 = 23.055$, $\alpha = 0.001$), and between control and treatment
populations ($\chi^2_2 = 4.726$, $\alpha = 0.094$). Mean abundance among control
sites was estimated to be $\hat{\mu}_c = 80.22$ and among treatment sites, $\hat{\mu}_T = 118.35$. Using test statistic (5.3)

$$d = \frac{4.37490 - 4.75878}{\sqrt{\dfrac{0.0259474}{4} + \dfrac{0.037305}{4}}} = -3.0527$$

mean abundance at control sites was found to be significantly less than at
treatment sites [$P(t_6 < -3.0527) = 0.011$]. As in the previous example,
the magnitude of the treatment effect was sufficiently large to be detected
with only four replicate sites. In many experiments, greater levels of replication will be needed in order to detect effects of smaller magnitude or in
the presence of greater environmental variation.

Posttest Analysis

Postdata Probability of Significance   The d_6-test statistic comparing mean
abundance of the least chipmunk at control and treatment sites in the chaining example above was found to be significant at $\alpha = 0.011$. This test

statistic is a ratio of two estimates; in the numerator is the estimated mean difference $\bar{X}_{C.} - \bar{X}_{T.}$ and in the denominator is the estimated standard error of this estimated mean difference. An excessively large value of the ratio ($|d_6| = 3.0527$) is generally explained as an excessively large numerator, but alternative explanations may be plausible because both numerator and denominator are subject to sampling error.

Because we are rejecting a null hypothesis concerning the numerator only, but can only assess the numerator relative to the denominator, we might therefore ask: If the ratio of numerator to denominator is 3.0527, and if the null hypothesis were true, what is the probability that the numerator exceeds its own critical value at the 0.05 level? The answer is read (interpolated) from the chi-square tables by entering these tables with $6 + 1 = 7$ degrees of freedom (add 1 to the degrees of freedom of d_6) with a chi-square value of

$$\left(\frac{6}{(3.0527)^2} + 1\right)(3.841) = 6.3140$$

and finding that the probability of a greater value is approximately 0.504 [i.e., $P(\chi_7^2 > 6.3140) = 0.504$]. Thus, given that $|d_6| = 3.0527$, then if the null hypothesis is true, the probability that the numerator of d_6 is significantly large (exceeding its 0.05 critical value) is only 0.504. The evidence for a significantly larger numerator is thus not very compelling, and this calculation casts a shadow on the nominal $\alpha = 0.011$ significance level for $d_6 = -3.0527$.

This calculation of postdata probability of a significantly large numerator in a two-sided t-test has the general form

$$P \left\{ \begin{array}{l} \text{Numerator is significant} \\ \text{at the } \alpha = 0.05 \text{ level,} \\ \text{given the realized value} \\ \text{of a t-statistic with } r \\ \text{degrees of freedom} \end{array} \right\} = P \left\{ \begin{array}{l} \text{Chi-square with} \\ r + 1 \text{ degrees} \\ \text{of freedom} \\ \text{exceeds} \\ \text{the value} \end{array} \left(\frac{r}{t^2} + 1\right)(3.841) \right\}. \quad (5.8)$$

The number 3.841 is the 5% critical value of chi-square with 1 degree of freedom, $3.841 = (1.96)^2$.

A similar calculation applies, more generally, to the F-test for simultaneously comparing more than two treatments. If the equality of v treatment means is tested by an F-test with $v - 1$ degrees of freedom in the numerator and ω degrees of freedom in the denominator, then, given the realized value (F) of this test statistic, the probability that the numerator of F exceeds its 0.05 critical value is read (interpolated) from a chi-square table, entering the table with $\omega + v - 1$ degrees of freedom and a chi-square value equal to

$$\left(\frac{\omega}{(v-1)F} + 1 \right) \chi^2_{v-1:\,.05} \tag{5.9}$$

where $\chi^2_{v-1:\,.05}$ denotes the 5% significant value of chi-square on $v - 1$ degrees of freedom. For example, if $v = 3$ treatments are being tested by an F-test with $\omega = 10$ degrees of freedom in the denominator, and the resulting F-statistic on 2 and 10 degrees of freedom happens to be $F = 4.10$, then

$$\left(\frac{\omega}{(v-1)F} + 1 \right) \chi^2_{2:\,.05} = \left(\frac{10}{2(4.10)} + 1 \right)(5.99) = 13.29.$$

Entering the chi-square table with $\omega + v - 1 = 10 + 3 - 1 = 12$ degrees of freedom, the value 13.29 is found (by linear interpolation) to have a P value of 0.348 [i.e., $P(\chi^2_{12} > 13.29) = 0.348$]. While the F-tables would say that $F_{2.10} = 4.10$ is significant at the 1% level, the chi-square tables would say that there is a 34.8% chance that the **numerator** of this F-statistic is significant at the 5% level.

These formulas [(5.8) and (5.9)] produce rather disconcerting results. In the case of the least chipmunk data where the predata probability of observing $|t| > 3.0527$ with 6 degrees of freedom is only 0.011, the conditional probability of a significant numerator is 0.504. With only 6 degrees of freedom available to estimate experimental error variance, however, there is clearly room for doubt concerning the reliability of a test, and indeed, this formula reveals that the conditional probability of a significantly large ($P < 0.05$) numerator can be at most $0.79 = P\{\chi^2_7 > 3.841\}$ no matter how large the observed $|t|$ value with 6 degrees of freedom.

These significance levels more honestly reflect the sensitivity of the test than power curves because in addition to measuring rejection rate of the null hypothesis, H_0, due to an excessively large numerator, the power is deceptively inflated by rejections resulting from an excessively small denominator. As degrees of freedom increase, this unsettling property of the t-test gradually vanishes. With 20 degrees of freedom, for example, and an observed $|t|$ value at exactly the 1% level in the t-table, one can be 90% confident that the numerator exceeds its 5% critical value; thus, with 20 degrees of freedom and nominal significance at exactly the 1% level, there is believable significance at the 5% level. The serious implications of these calculations to the interpretation of wildlife investigations with typically low levels of replication is apparent.

Estimation of Treatment Contrasts   After the tests of hypotheses have been completed, and if the null hypothesis (5.1) had been rejected, there may be interest in estimating the extent of change in animal abundance that has occurred as the result of treatment effects. In the two-treatment scenario, the contrast of interest is

$$\ln \mu_C - \ln \mu_T = \ln\left(\frac{\mu_C}{\mu_T}\right) \tag{5.10}$$

which measures the relative difference in mean abundance among the treatments. Numerators of the test statistics (5.3) and (5.6) can be used to estimate (5.10) and by equating d in (5.3) or (5.6) to a value of $t_{1-(\alpha/2)}$ appropriate for a two-tailed test, a $(1 - \alpha)100\%$ confidence interval estimate of (5.10) can be constructed. Because the null hypothesis has been rejected, it is likely that treatment variances will be unequal and consequently, an adjusted t' value (5.7) may be needed in confidence interval construction. Algebraic use of the inverse ln-transformation (i.e., e^x) on that interval estimate can then provide a confidence interval estimate of μ_C/μ_T, the fractional change in abundance as a result of treatment. Similarly, the numerator of (5.4) can be seen to provide an estimate of the linear contrast

$$\Delta = \mu_C - \mu_T \tag{5.11}$$

and the Z-statistic can be inverted to provide a confidence interval estimate of the numerical change in abundance as the result of the treatment effects. Furthermore, test statistic (5.5) can be inverted to provide a confidence interval estimate of μ_C/μ_T.

In constructing treatment contrasts, the reader is reminded that unbiased treatment means become biased when arranged in rank order. In other words, by placing the connotation that as a result of the treatment effects, treatment A can be expected to result in an *increase* of X animals over treatment B, the estimate of change is no longer unbiased. The two-tailed normal Z-test, say, $|Z| = |\hat{D}|/\sigma_D$, where $\hat{D} = \hat{\mu}_C - \hat{\mu}_T$ is commonly used to test the null hypothesis that the expected value of the difference in treatment means is zero, i.e., $H_0: E(\hat{D}) = 0$, against the alternative $H_a: E(\hat{D}) \neq 0$ in circumstances where the sign of $\Delta = E(\hat{D})$ is unknown. If the observed $|Z|$ exceeds $z_{1-(\alpha/2)}$, then H_0 is rejected in favor of H_a. The statistical inference does not ordinarily stop at this point but goes on to note the sign of the realized $\hat{D}$ and to infer that this same sign also holds for Δ. If $\hat{D} = \hat{\mu}_C - \hat{\mu}_T$ is found to be significantly different from zero, then the sign of this difference is usually of interest and is ascribed to Δ as an integral part of statistical inference.

The conclusion that $\text{sign}(\hat{D}) = \text{sign}(\Delta)$ is not a logical consequence of the rejection of H_o by a two-tailed $|Z|$-test and cannot be legitimately associated with the Type I error probability level α of this test. However, an inference such as $\mu_C > \mu_T$ is a statement concerning the location of the parameter Δ on the real line and might therefore appear to be associated with a confidence interval statement and a confidence level of $1 - \alpha$. Thus, if the confidence limits $\hat{D} \pm Z_{1-(\alpha/2)}\sigma_D$ do not bracket zero, then the assertion that $\text{sign}(\Delta) = \text{sign}(\hat{D})$ might appear to be sanctioned by the interval estimation procedure at a confidence level of $1 - \alpha$.

The fallacy is perhaps more subtle in this latter guise, which differs only semantically from the hypothesis testing situation. When the confidence interval lies entirely to one side of a fixed point (zero), e.g., to the right of zero, the fallacious line of reasoning might proceed as follows: I am 95% confident ($\alpha = 0.05$) that this interval contains the true Δ, and because all candidate Δ values within this interval are positive, I am 95% confident the true Δ is positive. Since no such confidence in the sign of Δ would have been expressed if the confidence limits had bracketed zero, this rationale is tantamount to hypothesis testing. Regardless of the semantics, the appeal for justification in level of confidence is an appeal to the operating characteristics of a statistical procedure that controls the long-run frequency of committing errors of a specific type.

The beforehand probability that $(\hat{D} - Z_{1-(\alpha/2)}\sigma_D, \hat{D} + Z_{1-(\alpha/2)}\sigma_D)$ will include Δ is $1 - \alpha$ whatever the numerical value of Δ, thereby justifying the $1 - \alpha$ level of confidence in the interval estimation procedure. The beforehand probability that $\text{sign}(\hat{D}) = \text{sign}(\Delta)$, however, is a function of the unknown true value of $|\Delta| = \delta\sigma_D$,

$$P\{\text{sign}(\hat{D}) = \text{sign}(\Delta)\} = \phi(\delta),$$

and the beforehand probability of the joint event that zero is not in the confidence interval *and* the sign of $\hat{D}$ agrees with the sign of Δ reduces to

$P\{0$ is not within the confidence interval

$$(\hat{D} - Z_{1-(\alpha/2)}\sigma_D, \hat{D} + Z_{1-(\alpha/2)}\sigma_D) \text{ and } \text{sign}(\hat{D}) = \text{sign}(\Delta)\} = \phi(\delta - Z_{1-(\alpha/2)})$$

$$(5.12)$$

which, likewise, depends on the unknown δ.

With the estimate $\hat{D} = (\hat{\mu}_C - \hat{\mu}_T)$ in hand, the issue of correctness of $\text{sign}(\hat{D})$ is addressed by the conditional probability

$$P\left\{\text{sign}(\hat{D}) = \text{sign}(\Delta) \left| \frac{|\hat{D}|}{\sigma_D} = d\right.\right\} = \frac{1}{1 + e^{-2\delta d}} \underset{[\text{def}]}{\equiv} L(2\delta d). \quad (5.13)$$

Although this logistic probability (Feller 1971, pp. 452–453) does depend on the unknown $\delta = |\Delta|/\sigma_D$, as well as on the observed value of $|\hat{D}|$, there is potential utility in the knowledge that if $|\Delta|$ exceeds a specified indifference value $|\Delta^*|$, then the probability of having correctly determined the sign of Δ is at least $L(2\delta^*d)$. Here $|\Delta^*|$ is a magnitude of treatment difference considered important to detect irrespective of Δ and $\delta^* = (|\Delta^*|/\sigma_D)$.

As a final caution when estimating treatment contrasts, although tests of effects and estimates of the ratios (μ_C/μ_T) do not necessarily require unbiased estimates of abundance in the presence of interpopulation homogeneity, estimation of the linear contrast $\mu_C - \mu_T$ does require valid estimates of abundance. If the primary objective of a manipulative experiment is estimation of differences in the mean abundance rather than strictly hypothesis testing, selection of survey models that permit intrapopulation heterogeneity (e.g., Otis et al. 1978) is essential. For this reason, careful consideration of the intended use of test results is important before implementing a manipulative experiment.

Designing a Completely Randomized Experiment

In designing a field experiment, the focus will be on choosing a design with sufficient power $(1 - \beta)$ to reject the null hypothesis (5.1) at a significance level of α, when the alternative is true with a magnitude $\Delta = |\mu_T - \mu_C|$. These quantitative objectives (α, β, Δ) are subjective criteria specified by an investigator that define the desired performance of an experiment. Levels of field replication and trapping effort at study plots will depend on the specification of these criteria, which, in turn, should reflect intent of the study. Preliminary investigations and screening studies may be based on large values of Δ and α for fixed β, whereas confirmatory studies will likely be designed with smaller values of Δ and α.

Power calculations $(1 - \beta)$ will be used to evaluate anticipated performance of various field designs with an objective of finding a design configuration that will provide sufficient power to detect a magnitude of treatment effect (Δ) considered important. Power of a completely randomized design to detect treatment differences will depend on the following: (1) test statistic employed; (2) α-level; (3) level of field replication (l_C, l_T); (4) capture probabilities (p_{1ij}, p_{2ij}, $i = C$, T and $j = 1, \ldots, l_i$); (5) magnitude of treatment difference, $|\mu_C - \mu_T|$; and (6) magnitude of plot-to-plot variances, σ_C^2 and σ_T^2. Analytic approximations for calculating power of tests are presented for cases 1–3. A prudent investigator may wish to consider only case 1 when there exists the possibility of interpopulation heterogeneity in capture rates. Cases 2 and 3 are then restricted to the situation where previous studies reveal a consistent pattern of interpopulation homogeneity. Alternatively, power calculations under cases 2 and 3 may be used to determine the maximum performance of a manipulative experiment under favorable test conditions.

Power Calculations for Case 1   For the case when capture probabilities are heterogeneous, both within and between treatments for a Lincoln Index, power of the test can be estimated from the noncentrality parameter

$$\phi = \frac{1}{\sqrt{2}}$$

$$\cdot \frac{|\ln \mu_C - \ln \mu_T|}{\sqrt{\frac{1}{l_C}\left(\frac{1}{\mu_C} + \frac{\sigma_C^2}{\mu_C^3}\right)^2\left(\sigma_C^2 + \frac{\mu_C}{l_C}\sum_{j=1}^{l_C}\frac{q_{1j}q_{2j}}{p_{1j}p_{2j}}\right) + \frac{1}{l_T}\left(\frac{1}{\mu_T} + \frac{\sigma_T^2}{\mu_T^3}\right)^2\left(\sigma_T^2 + \frac{\mu_T}{l_T}\sum_{j=1}^{l_T}\frac{q_{1j}q_{2j}}{p_{1j}p_{2j}}\right)}}$$

(5.14)

for a noncentral F-distribution with 1 and $l_T + l_C - 2$ degrees of freedom and where μ_T is the mean of treatment populations; μ_C, the mean of control populations; σ_T^2, spatial variance among treatment populations; and σ_C^2, spatial variance among control populations. The noncentrality parameter ϕ corresponds to the d-statistic (5.3). Tables for noncentral F-distribution (Appendixes 2 and 3) and ϕ can be used to calculate power for a two-tailed test. To calculate the power of a one-tailed test with significance level α, use tables at 2α-level and interpret it as α to find the proper power estimate (Appendixes 2 and 3).

Typically, the specific pattern of interpopulation heterogeneity will be unknown and power calculations will have to be based on an average anticipated per-period capture probability, $\bar{p}$. As such, (5.14) simplifies to the expression

$$\phi = \frac{1}{\sqrt{2}}$$

$$\cdot \frac{|\ln \mu_C - \ln \mu_T|}{\sqrt{\frac{1}{l_C}\left(\frac{1}{\mu_C} + \frac{\sigma_C^2}{\mu_C^3}\right)^2\left[\sigma_C^2 + \mu_C\left(\frac{1}{\bar{p}} - 1\right)^2\right] + \frac{1}{l_T}\left(\frac{1}{\mu_T} + \frac{\sigma_T^2}{\mu_T^3}\right)^2\left[\sigma_T^2 + \mu_T\left(\frac{1}{\bar{p}} - 1\right)^2\right]}}$$

(5.15)

The approximation obtained by substituting an average $\bar{p}$ value for capture probabilities may not give as reliable results as (5.14), but little else can be done without prior information on individual rates of capture for each population. What (5.15) does accomplish is to maintain the sampling error structure that is characteristic of using absolute abundance estimates in tests of treatment effects (5.3).

Power Calculations for Case 2   The power of a two-tailed test using Z-statistic (5.4) can be estimated from the expression

$$Z_\beta = \frac{|\mu_C - \mu_T|}{\sqrt{\frac{1}{l_C}\left(\frac{\mu_C q_1 q_2}{p_1 p_2} + \sigma_C^2\right) + \frac{1}{l_T}\left(\frac{\mu_T q_1 q_2}{p_1 p_2} + \sigma_T^2\right)}} - Z_\alpha,$$

(5.16)

where Z_α represents the value of the standard normal deviate corresponding to $P(Z < |Z_\alpha|) = 1 - \alpha$ (where α is the significance level of the test) and where Z_β is a standard normal deviate.

Power is computed from the cumulative standard normal distribution from the expression

$$\text{Power} = 1 - P(Z > Z_\beta) = P(Z < Z_\beta).$$

Again, a common average per-period capture rate, $\bar{p}$, may have to be used in power calculations in the absence of specific information to the contrary. In those circumstances, (5.16) reduces to

$$Z_\beta = \frac{|\mu_C - \mu_T|}{\sqrt{\dfrac{1}{l_C}\left[\mu_C\left(\dfrac{1}{\bar{p}} - 1\right)^2 + \sigma_{\tilde{C}}^2\right] + \dfrac{1}{l_T}\left[\mu_T\left(\dfrac{1}{\bar{p}} - 1\right)^2 + \sigma_{\tilde{T}}^2\right]}} - Z_\alpha. \qquad (5.17)$$

In using either (5.16) or (5.17), the same warning as to $l_C = l_T > 2$ pertain in the power calculations as in the use of (5.4) in analysis of capture data. For a one-tailed test under $H_a\colon \mu_C > \mu_T$, power is computed from $P(Z \le Z_\beta)$, where Z_α is a standard normal deviate corresponding to $P(Z > Z_\alpha) = \alpha$. Power is computed similarly under the alternative $H_a\colon \mu_C < \mu_T$.

Power Calculations for Case 3   Because test statistics (5.3) and (5.6) are similar in form except for the use of absolute abundance ($\hat{N}$) and catch index (r) data, respectively, power calculations are similar as well; the difference is in the expressions used for experimental error. Analytic approximation for the power of case 3 test statistic uses the noncentrality parameter

$$\phi = \frac{1}{\sqrt{2}} \cdot \frac{|\ln \mu_C - \ln \mu_T|}{\sqrt{\dfrac{1}{l_C}\left[\dfrac{q_1 q_2}{\mu_C(1 - q_1 q_2)} + \dfrac{\sigma_{\tilde{C}}^2}{\mu_{\tilde{C}}^2}\right] + \dfrac{1}{l_T}\left[\dfrac{q_1 q_2}{\mu_T(1 - q_1 q_2)} + \dfrac{\sigma_{\tilde{T}}^2}{\mu_{\tilde{T}}^2}\right]}} \qquad (5.18)$$

for a noncentral F with 1 and $(l_C + l_T - 2)$ degrees of freedom. With the value of ϕ, the power of the test $(1 - \beta)$ can be found by entry into noncentral F-tables (Appendixes 2 and 3).

EXAMPLE   The manipulative experiment considered consists of twelve 6-ha study plots distributed in pinyon-juniper habitat. A random selection of six plots will be chained and its effects on least chipmunk (*Eutamias minimus*) abundance assessed 2 years following application of treatment. Controls will be assumed to have a mean abundance of 54 individuals (9 ha^{-1}) per plot while the treated areas will have 108 individuals (18 ha^{-1}) per site (Table 5.5).

To progress with the power calculations, "guesstimates" of anticipated plot-to-plot variance must be obtained (Table 5.5). The large difference in mean abundance suggests that plot-

Table 5.5

Parameters Used in Designing an Experiment
to Assess the Effects of Chaining on
Abundance of Least Chipmunks

Parameter	Controls	Chained
$l_C = l_T$	6	6
A	6 ha	6 ha
μ_N	54	108
σ_N^2	140.71	217.75
$\bar{p}_1$	0.2	0.2
$\bar{p}_2$	0.2	0.2

to-plot variances will also differ between control and chained sites. Preferably, preliminary surveys might provide estimates of σ_C^2 and σ_T^2. In the absence of such studies, mean–variance relationship (2.22) will be used. From (2.22), the values of $\sigma_C^2 = 140.71$ and $\sigma_T^2 = 217.75$ are obtained.

Using power calculations under case 1 (5.15), the noncentrality parameter for a noncentral F with 1 and 10 degrees of freedom and an average per-period capture rate of $\bar{p} = 0.2$ is computed to be

$$\phi = \frac{1}{\sqrt{2}}$$

$$\cdot \frac{|\ln(54) - \ln(108)|}{\sqrt{\frac{1}{6}\left(\frac{1}{54} + \frac{140.71}{(54)^3}\right)^2\left[140.71 + 54\left(\frac{1}{0.2} - 1\right)\right]^2 + \frac{1}{6}\left(\frac{1}{108} + \frac{217.75}{(108)^3}\right)^2\left[217.75 + 108\left(\frac{1}{0.2} - 1\right)\right]^2}}$$

or $\phi = 1.6163$. From the tables in Appendix 2, the proposed study has a projected power of $1 - \beta = 0.685$ at $\alpha = 0.10$ or $1 - \beta = 0.541$ at 0.05. Because projected power of this experiment with $l = 6$ and $\bar{p} = 0.2$ is rather low, a more detailed analysis with various combinations of l and $\bar{p}$ is indicated (Figure 5.5).

The power curves (Figure 5.5) indicate design configurations with $l = 6$, $\bar{p} = 0.25$, or $l = 4$, $\bar{p} = 0.3$ or still $l = 2$ and $\bar{p} = 0.5$ will result in a manipulative experiment with $1 - \beta > 0.90$. Although any of these designs may be equally effective, costs may be markedly different. The section on design optimization will illustrate how economic considerations can be used to select a design configuration that is both effective and efficient. Note that if interpopulation homogeneity can be assumed, the power of the test is substantially increased for at $\alpha = 0.10$ and $l = 2$, $\bar{p} = 0.2$, the projected power with (5.18) is $1 - \beta = 0.847(\phi = 2.0325)$ compared to $1 - \beta = 0.347$ with (5.15).

Robustness of Power Calculations to Intrapopulation Heterogeneity   The assumption of intrapopulation homogeneity of capture probabilities will have an effect on accuracy of projected power of statistical tests in the presence of interpopulation heterogeneity. In performing power calculations, an average per-period capture probability ($\bar{p}$) among individuals must be used in the analytic approximations (5.15), (5.17), and (5.18). Computer simula-

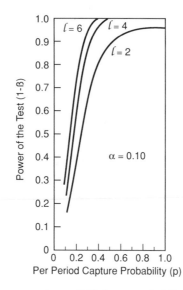

Figure 5.5 Power curves at $\alpha = 0.10$ for a manipulative experiment using a completely randomized design with various combinations of field replications ($l_C = l_T = l$) and per-period capture probabilities (p). Control populations assumed to have a mean abundance of $\mu_C = 54(\sigma_C^2 = 140.71)$ and treatment populations, a mean of $\mu_T = 108(\sigma_T^2 = 217.75)$.

tions indicate that such power calculations will likely underestimate $1 - \beta$ when heterogeneity exists and only the average $\bar{p}$ is used in projections (Table 5.6). The consequence of implementing design decisions based on (5.15), (5.17), and (5.18) will be to devise a field study with greater power than projected, assuming intrapopulation homogeneity. The simulation studies were conducted under conditions of intrapopulation heterogeneity in the presence of interpopulation homogeneity. As discussed earlier, when both intra- and interpopulation heterogeneity occur concurrently, survey models (Otis et al. 1978) appropriate for valid abundance estimation in presence of intrapopulation heterogeneity must be used in tests of hypotheses. The typically greater sampling error associated with these abundance estimators will have the effect of reducing the power of treatment comparisons.

Generalization of Power Calculations For survey models other than the Lincoln Index, the power of treatment comparisons can be computed using (5.14) under case 1 by substituting the average sampling error for the alternative mark–recapture estimator for that of the Lincoln Index. The generalization of the noncentrality parameter (5.14) for any mark–recapture survey can be written as

Table 5.6
Estimates of Power of the Test (5.1) for a Two-Treatment, Completely Randomized Design Using Statistic (5.3)*

(a)

Replication (l)	Analytic approximation (5.17)		Homogeneous p = 0.3		Heterogeneous ($\bar{p}$ = 0.3)					
					s = 0.6, λ = 1.4		s = 3, λ = 7		s = 0.42, λ = 0.98	
	α = 0.10	α = 0.05	α = 0.10	α = 0.05	α = 0.10	α = 0.05	α = 0.10	α = 0.05	α = 0.10	α = 0.05
2	0.195	0.104	0.226	0.119	0.236	0.122	0.226	0.116	0.260	0.127
4	0.395	0.247	0.439	0.303	0.548	0.392	0.467	0.317	0.550	0.382
6	0.513	0.371	0.566	0.437	0.716	0.581	0.625	0.485	0.743	0.611
8	0.629	0.490	0.697	0.551	0.828	0.732	0.731	0.612	0.865	0.761
10	0.721	0.594	0.773	0.660	0.903	0.832	0.836	0.734	0.916	0.842

(b)

Replication (l)	Analytic approximation (5.17)		Homogeneous p = 0.5		Heterogeneous ($\bar{p}$ = 0.5)					
					s = 0.5, λ = 0.5		s = 1, λ = 1		s = 3, λ = 3	
	α = 0.10	α = 0.05	α = 0.10	α = 0.05	α = 0.10	α = 0.05	α = 0.10	α = 0.05	α = 0.10	α = 0.05
2	0.291	0.159	0.467	0.287	0.313	0.159	0.318	0.161	0.316	0.164
4	0.626	0.466	0.648	0.483	0.700	0.572	0.681	0.532	0.671	0.512
6	0.811	0.692	0.818	0.703	0.863	0.759	0.862	0.751	0.848	0.750
8	0.916	0.832	0.924	0.847	0.941	0.886	0.939	0.884	0.928	0.868
10	0.957	0.900	0.969	0.934	0.975	0.946	0.974	0.946	0.973	0.937

* Based on 2000 computer simulations per scenario when interpopulation homogeneity exists and individual capture rates are homogeneous (p) or beta-distributed (s, λ) with expected value $E(p) = s/(s + \lambda)$. Animal abundance was negative-binomially distributed with $\mu_C = 50$, $\sigma_C^2 = 100$ and $\mu_T = 74$, $\sigma_T^2 = 222.6$. Analytic approximations to the power are based on (5.17) using an average anticipated capture rate $(\bar{p})$ of (a) $\bar{p} = 0.3$ or (b) $\bar{p} = 0.5$.

$$\phi = \frac{1}{\sqrt{2}} \cdot \frac{|\ln \mu_C - \ln \mu_T|}{\sqrt{\frac{1}{l_C}\left(\frac{1}{\mu_C} + \frac{\sigma_C^2}{\mu_C^3}\right)^2 (\sigma_C^2 + \overline{\text{Var}(\hat{N}_{Cj})}) + \frac{1}{l_T}\left(\frac{1}{\mu_T} + \frac{\sigma_T^2}{\mu_T^3}\right)^2 (\sigma_T^2 + \overline{\text{Var}(\hat{N}_{Tj})})}}$$

$$(5.19)$$

where $\overline{\text{Var}(\hat{N}_{ij})}$ represents the average sample variance in abundance estimation for the ith treatment ($i = C, T$) and where ϕ is associated with a noncentral F-distribution with 1 and $l_T + l_C - 2$ degrees of freedom. Power calculations based on (5.19) assume analysis using test statistic (5.3) based on ln-transformed $\hat{N}_{ij}$ values under the appropriate survey model.

Similarly, under case 3, the noncentrality parameter can be generalized for other mark–recapture techniques as

$$\phi = \frac{1}{\sqrt{2}} \cdot \frac{|\ln \mu_C - \ln \mu_T|}{\sqrt{\frac{1}{l_C}\left[\frac{\sigma_C^2}{\mu_C^2} + \frac{1-P}{\mu_C P}\right] + \frac{1}{l_T}\left[\frac{\sigma_T^2}{\mu_T^2} + \frac{1-P}{\mu_T P}\right]}} \qquad (5.20)$$

where P is the overall probability of capture of an animal during the course of the survey and ϕ is the noncentrality parameter for a noncentral F-distribution with 1 and $l_T + l_C - 2$ degrees of freedom. Power calculations based on (5.20) correspond to test statistic (5.6) based on the ln-transformed r_{ij} values and the assumption of approximate log-normality. For any survey model, the overall probability of capture (P) can always be estimated as

$$\hat{P} = \frac{r}{\hat{N}}$$

for some valid estimator $\hat{N}$ of N. In *post hoc* power calculations, a pooled estimator of P would be used where

$$\hat{P} = \frac{r_{..}}{\hat{N}_{..}}$$

and where $\hat{N}_{..}$ is the pooled estimate of abundance across the $l_C + l_T$ sites obtained by pooling the capture data from all plots.

Optimizing Randomized Designs
In the design of a multiplot population study, an investigator must make decisions as to number and size of study plots, trapping effort, and survey duration of the study. Each of these design parameters will have an influence on performance of the experiment. Review of small-mammal studies (Table 5.1) shows a wide variation in plot size (0.08–16.2-ha), trapping effort

(24.0–379.8 traps/ha), and level of replication (one to six plots per treatment) incorporated in manipulative experiments on the same species. Method sections of research papers rarely include the rationale used in making these design decisions or a detailed evaluation of performance of such studies.

Optimization procedures usually focus on minimizing experimental error in multiplot studies for fixed total study costs. However, we have chosen to focus on the power of tests because minimizing the experimental error does not guarantee that power $(1 - \beta)$ will be maximized. Power of a test depends not only on magnitude of experimental error variance but also on the degrees of freedom for the error term [in the d-statistics (5.3) and (5.6)]. Scheffé (1959, p. 236) mentions that optimizing tests remains an unsolved problem. Using a graphical approach, the task of optimizing power of a two-treatment test can be made tractable. When the primary focus of the experiment is to precisely estimate a treatment contrast, the more straightforward task of minimizing the experimental error variance should be performed.

The first step in optimization is construction of a cost function. In a manipulative experiment, size of the field study is determined by the number of replicate plots per treatment (l), area size of these study plots (A), trapping effort at each plot (f), and number of survey periods (d). Using these design criteria, Skalski (1985b) developed the multiplot cost function

$$C_0 = C_1 l + C_2\sqrt{l} + C_3 d\sqrt{l} + C_4 lf + C_5 ldf + C_6 ldAf$$
$$+ C_7 l\sqrt{Af} + C_8 l\sqrt{Af}, \qquad (5.21)$$

which can be used to optimize a manipulative experiment in conjunction with noncentrality parameters of the noncentral F-distributions (5.15) and (5.18). For the two-treatment, completely randomized design, optimization will be illustrated by determining optimal levels of plot replication (l) and trapping effort (f) of a Lincoln Index for fixed power, $(1 - \beta)$. By holding d and A fixed, a simpler cost function in terms of only l and f results, where

$$C_0 = C_1 l + C_2\sqrt{l} + C_3 lf + C_4 l\sqrt{f}. \qquad (5.22)$$

Solving (5.22) in terms of the number of plots (l),

$$l = \left[\frac{-C_2 + \sqrt{C_2^2 + 4C_0(C_1 + C_3 f + C_4\sqrt{f})}}{2(C_1 + C_3 f + C_4\sqrt{f})}\right]^2, \qquad (5.23)$$

cost contours can be graphed by successively substituting into (5.23) various values of f for different budgets C_0.

The next step in optimization is finding solutions to the power function for the test of effects in terms of l and f. Letting $l_C = l_T = l$ and using the catch–effort relationship, $p = 1 - e^{-df/A}$, the noncentrality parameter for case 1 becomes

$$\phi = \frac{1}{\sqrt{2}}$$

$$\cdot \frac{|\ln \mu_C - \ln \mu_T|}{\sqrt{\frac{1}{l}\left[\left(\frac{1}{\mu_C} + \frac{\sigma_C^2}{\mu_C^3}\right)^2 (\sigma_C^2 + \mu_C(e^{cf/A} - 1)^{-2}) + \left(\frac{1}{\mu_T} + \frac{\sigma_T^2}{\mu_T^3}\right)^2 (\sigma_T^2 + \mu_T(e^{cf/A} - 1)^{-2})\right]}}$$

(5.24)

Again, (5.24) can be solved in terms of one of the design criteria, for example,

$$f = \frac{A}{c} \ln \left[1 + \sqrt{\frac{\left[\left(\frac{1}{\mu_C} + \frac{\sigma_C^2}{\mu_C^3}\right)^2 \mu_C + \left(\frac{1}{\mu_T} + \frac{\sigma_T^2}{\mu_T^3}\right)^2 \mu_T\right]}{\left[\frac{l(\ln \mu_C - \ln \mu_T)^2}{2\phi^2}\right] - \left(\frac{1}{\mu_C} + \frac{\sigma_C^2}{\mu_C^3}\right)^2 \sigma_C^2 - \left(\frac{1}{\mu_T} + \frac{\sigma_T^2}{\mu_T^3}\right)^2 \sigma_T^2}}\right]$$

(5.25)

to readily construct power contours. Choosing an α-level and a value of $1 - \beta$, various values of l can be substituted into (5.25) to compute the corresponding values of f. To perform these calculations, an appropriate value of ϕ (Appendix 3) must be substituted into (5.25) with each successive choice of l. For l replicate study plots per treatment, ϕ corresponds to a noncentral F-distribution with 1 and $f_2 = 2l - 2$ degrees of freedom. The noncentral F-tables of Appendix 2 have been rearranged in Appendix 3 to facilitate reading the value of ϕ directly (or with simple interpolation) for $\alpha = 0.02, 0.05, 0.10,$ or 0.20 and $1 - \beta = 0.10 - 0.99$. By repeated choices of l, a power contour can be constructed from pairs of (l, f) at a specific combination of α and $1 - \beta$.

Through a process of trial and error, a cost contour that shares a single common tangent with the power contour can be found from (5.25). The point of intersection of these two curves indicates the optimal selection of l and f, which provides a power of $1 - \beta$ at α for a minimum budget of C_0.

Alternatively, if the purpose of design optimization is to maximize the power of the test for a fixed budget of C_0, the process of trial and error as given above is reversed. Using (5.23), a single cost contour is constructed and a series of power contours from (5.25) are plotted to find the curve corresponding to the greatest value of $1 - \beta$ that shares a common tangent with the cost contour. The point of intersection indicates the pair of values (l, f) that are predicted to provide maximum power $1 - \beta$ at α for fixed total costs C_0.

The advantage of the graphical approach presented for optimization is the ability to circumvent the need to differentiate across the noncentral

F-distribution in order to find the optimal design configuration. While tractable, numerical techniques would be required to obtain the same results that the graphical technique provides. Perhaps more compelling, the graphical technique by its very nature requires an investigator to seriously consider interrelationships between design configurations, study performance, and research costs. Given the current design of many wildlife studies (Table 5.1), such consideration may do much to help improve the performance of manipulative experiments.

When three or more of the design criteria defined by A, f, l, and d are involved in design optimization, numerical methods are necessary to optimize across (5.21) and (5.15), (5.17), or (5.18). A numerical approach is the only method to use if an exact solution to optimization is required. For many investigators and most situations, a pairwise use of the preceding graphical technique will provide sufficient insight to efficiently design a field study. Obviously, pairwise solutions will not lead to the same overall solution as the analysis of several parameters. Because there are $\binom{4}{2} = 6$ pairwise choices of A, f, l, and d, the necessary calculations are more than can be presented here. The reader should use procedures similar to those illustrated above to find the optimal design with other pairwise selections of the design parameters.

EXAMPLE   To illustrate optimization procedures, a two-treatment, completely randomized design will be considered for an experiment anticipated to show a reduction in mean animal abundance from 18 to 9 individuals per hectare under treatment conditions. Plot size will be treated as a constant at 3.25 ha ($\mu_C = 58.5$, $\mu_T = 29.25$), resulting in anticipated population variances of $\sigma_C^2 = 147.98$, $\sigma_T^2 = 95.62$ from (2.22). The objective of optimization will be to find the level of trapping effort (f) and field replication (l) that will minimize the total study cost (C_0) for the fixed power of the test of (5.1) at $1 - \beta = 0.8$ and $\alpha = 0.10$ ($\alpha = 0.05$ one-tailed test). Population surveys at each plot will be conducted using the single mark–recapture method with an anticipated catch coefficient of $c = 0.0325$.

Substituting population parameters, catch coefficient, and appropriate values of ϕ into (5.25), values of f are found for various levels of l. Plotting the paired values of l and f, a power contour can be drawn (solid line, Figure 5.6). The next step in the optimization process is to solve for l and f from an appropriate cost function for various total costs, C_0. Considering the cost of trapping devices and labor, the following cost function results:

$$C_0 = 40l + 22.38\sqrt{l} + 17.52lf + 131.57l\sqrt{f}. \tag{5.26}$$

By trial and error, a value of $C_0 = 9400$ is found for the cost contour (dashed line, Figure 5.6) that intersects the power contour (for $1 - \beta = 0.8$, $\alpha = 0.10$) at a single point. From the point of common tangent between cost and power contours, an optimal design configuration of $f = 60$ and $l = 4.5$ is found to minimize study costs ($C_0 = \$9400$) for the specified power of $1 - \beta = 0.8$, $\alpha = 0.10$ ($\alpha = 0.05$ one-tailed). Because 4.5 replicate plots per treatment cannot be implemented, the level of replication must be raised to 5 or lowered to 4 plots per treatment, with an appropriate adjustment in trapping effort (f). The design configuration that yields maximum power is the optimal selection.

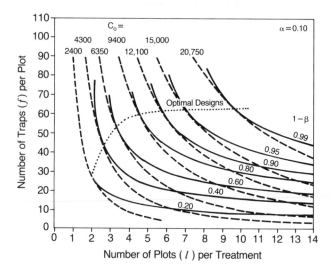

Figure 5.6 Power contours for the test of the treatment hypothesis [Eq. (5.1), solid lines] and solutions to the cost function [Eq. (5.26), dashed lines] as functions of trapping effort (f) and plot replication (l). Optimal design configurations sketched from identification of individual solutions (dotted line).

For still other values of $1 - \beta$, the search for optimal combinations of l and f can be conducted by the same process of trial and error using graphical techniques (Figure 5.6). By repeatedly finding solutions to an optimal design, a third curve can be drawn representing the optimal design curve. Along this curve are all combinations of l and f that optimize a design for various total costs, C_0, and levels of power, $1 - \beta$. Trapping effort (f) levels off at about 62 traps per plot once total available funds reach \$10,000 (Figure 5.6). Any additional funds over and above that budget go directly to establishment of more study plots. It is important to note that an allocation of 62 traps is not anticipated to result in complete enumeration of populations during the course of the surveys. The per-period capture probability is projected to be

$$p = 1 - e^{-0.325(62)/3.25} = 0.462,$$

and, during the survey, only 71% of the animals are expected to be caught at least once. Hence, the optimal design does not completely enumerate the study populations as so many investigators attempt (Table 5.1). Rather, optimal allocation is to increase replication as budgets increase, while sampling error is held at moderate levels. In all cases, minimal level of replication per treatment is $l = 2$.

Paired Treatment Experimental Design

Scenario
A paired treatment design is the simplest use of blocking in field studies. The experimental design consists of $2l$ study plots that are paired in geographically similar habitats to form l-blocks. By minimizing habitat differences, it is hoped that animal abundance will be similar within paired plots.

Effective pairing may be accomplished by locating plots in close physical proximity within habitat types. Close proximity may promote similar microhabitat conditions resulting in a positive correlation in abundance, and enhancing power of statistical tests to detect treatment effects. Within each pair of plots, control and treatment conditions are then randomly assigned to the areas. In practice, knowledge of local conditions and plant–animal associations will be important for proper pairing.

The process of pairing study plots should result in the majority of environmental variation being associated with differences between pairs. With this arrangement of sites, experimental error is reduced by identifying assignable causes for variation in animal abundance. Investigators may take advantage of this blocking by assigning treatment pairs to diverse habitat types. In so doing, inferences to treatment effects can be made to a range of environmental conditions without loss in precision.

When numerous control–treatment plots are to be surveyed during the course of the experiment, sites within a pair should be surveyed simultaneously to eliminate temporal effects on abundance. Other paired sites would then be sampled sequentially to complete the experimental design. Simultaneous surveys of paired sites enhance the prospects of homogeneous capture probabilities within pairs and allow an analysis based on catch indices. As in all experiments, selection of study sites should be a random sample (or stratified random sample) from the target population to which experimental results are to be inferred.

An alternative use of the paired design is in estimating temporal change in areal mean abundance between two sampling periods. By blocking on study plot location, statistical analysis can take advantage of the positive correlation in animal abundance through time to improve the precision of population comparisons. Although the effect of time from t to t_{i+1} cannot be randomized in this design, the study plots should be a random sample from the geographic region of inference. In this scenario, interest is not in hypothesis testing (assertion of no change can be rejected *a priori*) but rather in estimating a contrast for the difference in mean abundance through time.

The null hypothesis tested in paired designs is again that there is no difference in mean abundance between control and treatment conditions. Let K_i be the proportional abundance between control and treatment sites within a pair, $K_i = N_{C_i}/N_{T_i}$. The null hypothesis of no treatment effect can then be written as

$$H_0: \mu_K = 1 \tag{5.27}$$

against the alternative hypothesis of

$$H_a: \mu_K \neq 1.$$

If *a priori* knowledge exists about the potential treatment effect, one-sided hypotheses may be appropriate. Here again the use of proportional abundance (K) implies that treatment effects are expected to have a multiplicative influence on animal abundance.

Test Statistics

The test of the null hypothesis of no treatment effect (5.27) is performed on the basis of a statistical test of mean proportional abundance (μ_K). The model for the statistical analysis can be written as

$$\hat{K}_i = \mu \epsilon_i \tag{5.28}$$

where $\hat{K}_i$ is the ratio of faunal abundance estimates at control and treatment plots for the ith pair, $i = 1, \ldots, l$; μ is the mean ratio for control–treatment pairs; and ϵ_i is the random error term $(i = 1, \ldots, l)$. When populations in the experiment are censused using the single-mark–recapture method, an ln-transformation of the K is appropriate, and the resulting ln-linear model equation can be written in the form

$$\ln \hat{K}_i = \ln \mu + \ln \epsilon_i.$$

The form of the test statistic depends on assumptions concerning homogeneity of capture probabilities among populations in the experiment. There are only two alternative assumptions to consider:

Case 1: Capture probabilities for control and treatment populations within a pair are heterogeneous.

Case 2: Capture probabilities for control and treatment populations within a pair are homogeneous.

Test statistics are described below for the two experimental scenarios that may be encountered.

Tests of homogeneity (see Chapter 3) should be performed in identifying the types of interpopulation heterogeneity that may exist (i.e., cases 1, 2). Each pair of populations should be tested independently for the presence of heterogeneity. Proportional abundance should be estimated on the basis of absolute abundance (i.e., $\hat{K}_i = \hat{N}_{C_i}/\hat{N}_{T_i}$) if interpopulation heterogeneity is detected within paired plots. If interpopulation homogeneity can be substantiated, then proportional abundance should be estimated on the basis of catch indices (i.e., $\hat{K}_i = r_{C_i}/r_{T_i}$) for that pair of plots. The objective of the test of homogeneity is to identify the most appropriate estimator of proportional abundance that has minimal sampling variance on a pair-by-pair (pairwise) basis.

For experiments based on Lincoln Index surveys and case 1 heterogeneity, proportional abundance for a pair of plots is estimated as the ratio of Chapman (1951) estimators, where

$$\hat{K}_i = \frac{\dfrac{(n_{C1i} + 1)(n_{C2i} + 1)}{(m_{Ci} + 1)} - 1}{\dfrac{(n_{T1i} + 1)(n_{T2i} + 1)}{(m_{Ti} + 1)} - 1}. \qquad (5.29)$$

This estimate of K, while biased, results in a test statistic symmetric about 0 when testing the null hypothesis (5.27). For the other pairs of sites that show interpopulation homogeneity, proportional abundance should be estimated as $\hat{K}_i = r_{C_i}/r_{T_i}$ from the Lincoln Index data. Using the l estimates of K_i from the experiment, the statistic for testing the null hypothesis (5.27) can be written as

$$d = \frac{\dfrac{\sum\limits_{i=1}^{l} (\ln \hat{K}_i)}{l}}{\sqrt{\dfrac{S^2_{\ln \hat{K}_i}}{l}}} \qquad (5.30)$$

where

$$S^2_{\ln \hat{K}_i} = \frac{\sum\limits_{i=1}^{l} (\ln \hat{K}_i)^2 - \dfrac{\left(\sum\limits_{i=1}^{l} (\ln \hat{K}_i)\right)^2}{l}}{l - 1}$$

which is approximately distributed as a t-statistic with $(l - 1)$ degrees of freedom. Computer simulations show (5.30) to have nominal α-levels based on a Lincoln Index under both cases 1 and 2 (Skalski 1985a, pp. 196–211).

When population abundance is estimated using other mark–recapture models, the nature of test statistic (5.30) remains the same. Initially, tests of homogeneity should be performed to determine the most appropriate and efficient estimator of $K_i (i = 1, \ldots, l)$ for each pair of plots. In the absence of prior knowledge about the distributional properties of $\hat{K}$, and the belief that multiplicative effects on abundance are operating, ln-transformation of the data and the use of (5.30) is recommended.

Example Analysis

In assessing the possible effects of secondary poisoning of pesticide sprays on animal abundance, test conditions often can be randomly assigned to study plots. Capture data representing Lincoln Index surveys of Townsend's ground squirrels (*Spermophilus townsendii*) will be used to illustrate a field design to test effects of spraying programs at alfalfa fields on local small mammal populations (Table 5.7). Paired plots in five high abundance areas were identified for testing from a selection of potential sites in a county of interest. For each pair, two 5-ha study plots were established within a contiguous alfalfa field and located 0.4 km apart to preserve independence. The

Table 5.7

Capture Data, Abundance Estimates (3.5), and Catch Indices from Representative Lincoln
Index Surveys of Townsend's Ground Squirrel (*Spermophilus townsendii*) on Five Paired,
5.0-ha Plots under Control and Treated (Sprayed) Conditions*

| | Control | | | | | Treatment | | | | | Pairwise homogeneity | | |
Pair	n_1	n_2	m	r	N	n_1	n_2	m	r	N	χ_2^2	α	$K = r_C/r_T$
1	30	35	12	53	84.846	16	16	8	24	31.111	1.1445	0.564	2.208
2	98	26	23	101	110.375	57	12	10	59	67.545	0.7759	0.678	1.712
3	79	32	19	92	131.000	43	20	15	48	56.750	2.0349	0.362	1.917
4	54	35	16	73	115.471	28	19	9	38	57.000	0.0560	0.972	1.921
5	73	36	26	83	100.407	48	28	20	56	66.667	0.5960	0.742	1.482
											$\chi_{10}^2 = 4.6073$	0.916	

*Tests of pairwise homogeneity of capture probabilities (3.11) were performed for each
pair of control–treatment sites.

sprayed sites were randomly assigned within each pair and mark–recapture
surveys conducted 3 days after initial treatment. The purpose of this alleged
study was to determine short-term effects of the pesticide spraying program.
Three site pairs were treated and surveyed during 1 week followed by the
remaining two pairs the following week.

For each pair of control–treatment sites, chi-square tests of homogeneity (3.11) indicated no significant differences ($\alpha > 0.10$) in capture probabilities between paired sites (Table 5.7). The cumulative result of the five
separate pairwise tests also indicated no significant difference in capture
rates $[P(\chi_{10}^2 \geq 4.607) = 0.916]$. Using the marginals ($n_1.$, $n_2.$, $m.$) from
each pair of sites, a subsequent test of homogeneity indicated significant
differences in the pattern of captures among pairs of sites
$[P(\chi_8^2 \geq 76.985) \approx 0]$. From these test results, it appears that blocking had
a beneficial effect on homogeneity of capture probabilities, and the test of
the null hypothesis (5.27) should be performed using test statistic (5.30)
with relative abundance estimators $\hat{K}_i = r_{Ci}/r_{Ti}$. The test of the null hypothesis (5.27) resulted in a value $d_4 = 9.089$, which corresponds to a one-
tailed test at $\alpha = 0.004$. The conclusion drawn from the capture data was
that the pesticide had a significant effect, reducing short-term abundance of
ground squirrels at high-density alfalfa fields in the county where the experiment was performed.

Designing a Paired Experiment
As in the completely randomized design, the objective in designing a paired
experiment is to identify a design configuration with sufficient power
($1 - \beta$) to detect at a significance level α a treatment effect (Δ) considered

biologically important. The power of a paired treatment design will depend on the following conditions: (1) test statistic employed, (2) α-level, (3) level of field replication (l), (4) capture probabilities (p_{1ij}, p_{2ij}, $i = 1$, 2 and $j = 1, \ldots, l$), (5) magnitude of plot-to-plot variances σ_C^2 and σ_T^2, and (7) magnitude of correlation in abundance between control and treatment plots ρ_{N_C, N_T}. To calculate the power of a test of the null hypothesis (5.27), decisions about α and $|\mu_C - \mu_T|$, as well as estimates of spatial parameters (i.e., μ_C, μ_T, σ_C^2, σ_T^2, and ρ_{N_C, N_T}) and capture probabilities, must be available concerning the proposed experiment. A well-defined set of objectives along with preliminary survey data often will provide the necessary information to quantitatively design the field study.

Power Calculations for Case 1   The noncentrality parameter for the d-statistic (5.30)-based Lincoln Index surveys and $\hat{K}_i = \hat{N}_{Ci}/\hat{N}_{Ti}$ (for all $i = 1, \ldots, l$) can be written as

$$\phi = \frac{\sqrt{l}\,|\ln \mu_K'|}{\sqrt{2\sigma_{EX}^2}}, \qquad (5.31)$$

where

$$\mu_K' = \frac{\mu_C}{\mu_T}\left(\frac{1 - e^{-\mu_C p_1 p_2}}{1 - e^{-\mu_T p_1 p_2}}\right), \qquad (5.32)$$

and where

$$\sigma_{EX}^2 = \frac{q_1 q_2}{p_1 p_2}\left[\left(\frac{1}{\mu_C} + \frac{\sigma_C^2}{\mu_C^3}\right) + \left(\frac{1}{\mu_T} + \frac{\sigma_T^2}{\mu_T^3}\right)\right] + \left[\sigma_C^2\left(\frac{1}{\mu_C} + \frac{\sigma_C^2}{\mu_C^3}\right)^2 + \sigma_T^2\left(\frac{1}{\mu_T} + \frac{\sigma_T^2}{\mu_T^3}\right)^2\right.$$
$$\left. - 2\rho_{N_C, N_T}\sqrt{\sigma_C^2 \sigma_T^2}\left(\frac{1}{\mu_C} + \frac{\sigma_C^2}{\mu_C^3}\right)\left(\frac{1}{\mu_T} + \frac{\sigma_T^2}{\mu_T^3}\right)\right] \qquad (5.33)$$

and can be used in conjunction with a noncentral F-distribution with 1 and ($l - 1$) degrees of freedom under H_a to determine the power of the proposed experiment. The value of ϕ (5.31) can be used for entry into the noncentral F-tables (Appendixes 2 and 3) to calculate the power $(1 - \beta)$ of the test.

To correct for bias in $\hat{K} = \hat{N}_C/\hat{N}_T$ when p is small for a Lincoln Index, the numerator of the noncentrality parameter has been adjusted for the bias that would be observed under such conditions. This bias in μ_K' is represented by the factor

$$\frac{1 - e^{-\mu_C p_1 p_2}}{1 - e^{-\mu_T p_1 p_2}}$$

in (5.32). Although case 1 is appropriate when interpopulation heterogeneity in capture probabilities exists between all $2l$ populations, the expression for experimental error, σ_{EX}^2 in (5.33), is simplified by including only a common value for p_1 and p_2. This simplification is necessary because the exact expression of the heterogeneity will be unknown and best guesses for p_1 and p_2 must be used. The variance (5.33), however, takes into account the sampling error associated with estimating absolute abundance using Lincoln Indices.

Power Calculations for Case 2   The noncentrality parameter for the d-statistic (5.30) based on $\hat{K}_i = r_{Ci}/r_{Ti}$ from Lincoln Index surveys can be written as

$$\phi = \frac{\sqrt{l}\,|\ln \mu_K|}{\sqrt{2\sigma_{EX}^2}}, \tag{5.34}$$

where

$$\sigma_{EX}^2 = \frac{q_1 q_2}{(1 - q_1 q_2)}\left[\left(\frac{1}{\mu_C} + \frac{\sigma_C^2}{\mu_C^3}\right) + \left(\frac{1}{\mu_T} + \frac{\sigma_T^2}{\mu_T^3}\right)\right] + \left[\sigma_C^2\left(\frac{1}{\mu_C} + \frac{\sigma_C^2}{\mu_C^3}\right)^2\right.$$
$$\left. + \sigma_T^2\left(\frac{1}{\mu_T} + \frac{\sigma_T^2}{\mu_T^3}\right)^2 - 2\rho_{NC,NT}\sqrt{\sigma_C^2\sigma_T^2}\left(\frac{1}{\mu_C} + \frac{\sigma_C^2}{\mu_C^3}\right)\left(\frac{1}{\mu_T} + \frac{\sigma_T^2}{\mu_T^3}\right)\right], \tag{5.35}$$

and can be used in conjunction with a noncentral F-distribution with 1 and $(l - 1)$ degrees of freedom. Tables in Appendixes 2 and 3 can be used to determine the power of the test $(1 - \beta)$ from ϕ. While capture probabilities may vary from pair to pair in case 2, a common set of values $(p_1 = 1 - q_1$ and $p_2 = 1 - q_2)$ is used in the power calculations presented.

EXAMPLE   From the earlier discussion of a randomized design, power calculations indicated that six replicate 6-ha study plots in each of control and chained conditions resulted in a power of $1 - \beta = 0.685$ at $\alpha = 0.10$ when capture probabilities were $\bar{p} = 0.2$ ($\mu_C = 54$, $\sigma_C^2 = 140.71$, $\mu_T = 108$, and $\sigma_T^2 = 217.75$) for least chipmunk populations (Table 5.5). Similar calculations can be performed for a paired design with $l = 6$ and using the same spatial distributions under H_a to determine the advantage, if any, of pairing. The noncentrality parameter under case 1 for a paired design has the calculated value of

$$\phi = \frac{\sqrt{6}\,|\ln(0.4483)|}{\sqrt{2(0.5147)}} \approx 1.9370$$

which follows from

$$\mu_K' = \frac{54[1 - e^{-54(0.2)^2}]}{108[1 - e^{-108(0.2)^2}]} \approx 0.4483$$

and where

$$\sigma_{EX}^2 = \frac{0.8^2}{0.2^2}\left[\left(\frac{1}{54} + \frac{140.71}{54^3}\right) + \left(\frac{1}{108} + \frac{217.75}{108^3}\right)\right] + 140.71\left(\frac{1}{54} + \frac{140.71}{54^3}\right)^2$$

$$+ 217.75\left(\frac{1}{108} + \frac{217.75}{108^3}\right)^2$$

$$- 2(0.3)\sqrt{140.71 \cdot 217.75}\left(\frac{1}{54} + \frac{140.71}{54^3}\right)\left(\frac{1}{108} + \frac{217.75}{108^3}\right)$$

$$\approx 0.5147,$$

where $\rho_{(N_C, N_T)} = 0.3$. The value ρ_{N_C, N_T} is the correlation in abundance between the study sites within a pair. Using the tables in Appendix 2, a noncentrality parameter with a value of $\phi = 1.9370$ for a noncentral F-statistic with 1 and 5 degrees of freedom corresponds to a projected power of $1 - \beta = 0.752$ at $\alpha = 0.10$ or $1 - \beta = 0.588$ at $\alpha = 0.05$ for a two-tailed test. Since the chaining is anticipated to have a beneficial effect on chipmunk abundance, a one-tailed test is appropriate, in which case an $\alpha = 0.05$ test has a power of $1 - \beta = 0.752$. In order to compute the power for a one-tailed test, use (5.31) or (5.34) and compute the power for a two-tailed test at α and interpret at $\alpha/2$ (e.g., the power of a two-tailed test at $\alpha = 0.10$ is also the power of a one-tailed test at $\alpha = 0.05$). Had we anticipated a correlation of $\rho_{N_C, N_T} = 0.6$, the power of a one-tailed test at $\alpha = 0.05$ would be $1 - \beta = 0.766(\phi = 1.9742)$.

Generalization of Power Calculations For paired experiments performed with other than a Lincoln Index, a general form of the noncentrality parameter can be written for test statistic (5.30). However, the following expressions are not corrected for any biases that may be associated with the abundance estimators and as such, should be used for approximate use only. Under case 1, where all populations are quantified using absolute abundance estimators, the noncentrality parameter can be written as

$$\phi = \frac{\sqrt{l}\,|\ln \mu_K|}{\sqrt{2\sigma_{EX}^2}} \tag{5.36}$$

where

$$\sigma_{EX}^2 = \left(\frac{1}{\mu_C} + \frac{\sigma_{\hat{C}}^2}{\mu_{\hat{C}}^3}\right)^2 [\sigma_{\hat{C}}^2 + \overline{\mathrm{Var}(\hat{N}_{Cj})}] + \left(\frac{1}{\mu_T} + \frac{\sigma_{\hat{T}}^2}{\mu_{\hat{T}}^3}\right)^2 [\sigma_{\hat{T}}^2 + \overline{\mathrm{Var}(\hat{N}_{Tj})}]$$

$$- 2\rho_{N_C, N_T}\sqrt{\sigma_{\hat{C}}^2\sigma_{\hat{T}}^2}\left(\frac{1}{\mu_C} + \frac{\sigma_{\hat{C}}^2}{\mu_{\hat{C}}^3}\right)\left(\frac{1}{\mu_T} + \frac{\sigma_{\hat{T}}^2}{\mu_{\hat{T}}^3}\right) \tag{5.37}$$

and corresponds to a noncentral F-distribution with 1 and $(l - 1)$ degrees of freedom under H_a. Again, $\mathrm{Var}(\hat{N}_{ij})$ refers to the average sampling variance for abundance estimation in treatment $i = C, T$.

When capture probabilities can be assumed homogeneous within paired plots, the general form for the noncentrality parameter corresponding to test statistic (5.30) under case 2 is again (5.36) but instead of variance expression (5.37), use

$$\sigma_{\text{EX}}^2 = \left[\frac{\sigma_{\hat{C}}^2}{\mu_{\hat{C}}^2} + \frac{1-P}{\mu_C P} \right] + \left[\frac{\sigma_{\hat{T}}^2}{\mu_{\hat{T}}^2} + \frac{1-P}{\mu_T P} \right] - \frac{2\rho_{\mu_C, \mu_T} \sqrt{\sigma_{\hat{C}}^2 \sigma_{\hat{T}}^2}}{\mu_C \mu_T}$$

then the catch indices (i.e., r_{ij} values) can be assumed approximately lognormally distributed. Here P is the overall capture probability.

Optimizing Paired Designs

A graphical technique similar to the one used in optimizing a completely randomized design can be applied to (5.31) or (5.34) and cost function (5.21), to optimize a paired experimental design. The only noteworthy difference is the need to have an additional estimate of the correlation in abundance between control and treatment sites ρ_{N_C, N_T}. Preliminary survey data may provide the necessary estimate.

Open-Population Investigations

Open populations subject to immigration, emigration, recruitment, and mortality are the norm in wildlife investigations; use of closed-population models therefore approximates only the short-duration realities of nature. Indeed, all mark–recapture models, whether open or closed, are at best approximations of reality mandated by the limitations of parameter estimation. The art in wildlife population investigations is the acumen to define and estimate population parameters, and assess effects effectively through the combined efforts of field logistics and statistical modeling. No better an example of the importance of the synergism between field logistics and statistics exists than in the study of open populations. In previous discussions, the design and analysis of wildlife investigations has focused on the use of closed-population models. In this final section, we consider some of the unique considerations that apply to the study of open populations.

Utilizing Closed-Population Conditions

Statistical as well as logistical considerations strongly suggest using closed-population survey methods as often as possible when assessing effects on wild populations. Closed-population models may be a reasonable survey approach if the rate of population change (i.e., birth rate, mortality rate) is small relative to the initial population size and duration of study. Indeed, this scenario of using closed-population models is the actual situation in the best of survey conditions. Furthermore, the use of short-duration, intense sampling in conjunction with closed-population models has several advantages over the direct application of open-population surveys.

Mark–recapture models have difficulty differentiating intrapopulation heterogeneity in capture probabilities from concurrent recruitment and mortality processes. Cormack (1985) discussed the nonidentifiability of such effects when modeling open- and closed-population surveys. Because of the inherent difficulty of sorting out the effects of heterogeneity from recruitment and mortality using capture data, investigators must try to control one or both of these factors through survey design. Heterogeneity in trap response is always a possibility, and investigators have little control over its nature and extent. On the other hand, by designing population surveys over a relatively short duration, the influences of recruitment and mortality can be minimized. The flexibility in designing mark–recapture surveys can thus be best utilized in controlling lack of closure, leaving statistical modeling to address heterogeneity in trap response. Holding trapping effort constant during the course of the surveys may further simplify selection of an appropriate survey model by minimizing the prospects of nonestimable statistical models. Consequently, short-duration, intense surveys with constant effort provide the best circumstances for validly estimating animal abundance.

Pollock (1982) extends the idea of using short-duration, closed-population models in the study of open populations. He recommends a sequence of intense trapping sessions, each followed by a longer period of cessation of trapping, as an ideal survey design for open-population studies. Each session of intense trapping periods would be analyzed separately, using closed-population models to estimate animal abundance. The high capture probabilities over several to many days (≥ 10 days) provide ideal circumstances for characterizing trap response and selecting an appropriate closed-population model. Between trapping sessions, survival rates are estimated using various release–recapture models (e.g., Cormack 1964). The analysis of the release–recapture data provides survival estimates robust to heterogeneous capture probabilities. Use of short-duration, intense surveys in tests of population effects thus provides a useful approach for both open and closed populations.

Another reason for using closed-population models in open populations stems from the statistical requirement of independent observations in the tests of hypotheses (i.e., Z-, t-, or F-tests). Even among the study designs that use repeated population surveys over time [e.g., odd's ratio, CTP, and accident assessment], the abundance estimates or catch indices need to be conditionally independent [i.e., $\text{Cov}(\hat{N}_i, \hat{N}_j \mid N_i, N_j) = 0$]. However, repeated population estimates derived from an open-population model are correlated [i.e., $\text{Cov}(\hat{N}_i, \hat{N}_{i+1} \mid N_i, N_{i+1}) \neq 0$]. Consequently, repeated abundance estimates from a single open-population model should not be used in the tests of effects presented earlier in this chapter. Hence, long-duration, open-population surveys hold no distinct advantage over a sequence of independent, closed-population surveys.

Finally, looming above the statistical considerations encouraging the use of closed-population models are the economic and logistical constraints of a large multipopulation field investigation. Limitation of budget and labor often will not permit both intensive sampling of open populations and concurrently, extensive sampling of populations across time and landscape. Results of design optimality strongly suggest extensive sampling must be chosen over intensive surveys of a few open populations. The reason is that power of tests of hypotheses depend on the degrees of freedom associated with replicate plots treated alike as well as sampling error of the population surveys.

Using Open-Population Models

Open-population models such as Jolly–Seber (Jolly 1965; Seber 1965) or Manly–Parr (Manly and Parr 1968) provide the opportunity to investigate animal abundance along with the driving variables of mortality and recruitment. The auxiliary information on survival and births is valuable in helping to understand the population processes contributing to significant treatment effects on animal abundance. Treatment effects actually manifest themselves as direct effects on survival rates and recruitment. Animal abundance is then an integration of the treatment effects and any compensatory mechanisms operating in the population. It is therefore important to monitor shifts in abundance along with studies of survival and recruitment.

The desire to use open-population models and the joint study of abundance, survival, and recruitment must be tempered with the statistical realities of estimating numerous parameters with a single-survey model. The sampling errors of abundance estimates from open-population models will typically exceed those of closed-population models, all else being equal. For open-population surveys to be successful in multipopulation experiments, model selection techniques to identify parsimonious yet valid survey models is essential. Abundance estimates and estimates of sampling error from fitted open-population models can then be used in tests of hypotheses and power calculations presented in earlier chapters. Cormack (1985) presented a model selection technique based on fitting special cases of the Jolly–Seber (Jolly 1965; Seber 1965) model to open-population survey data.

Certain limitations exist when using open-population survey data in subsequent tests of effects on animal abundance. As mentioned earlier, the sequence of abundance estimates from an open-population analysis are correlated. Hence, only one of the abundance estimates from a population survey can be incorporated in tests of hypotheses. An *a priori* choice of which period will provide the abundance estimate ($\hat{N}_i$, $i = 2, \ldots, k - 1$) for inclusion in hypothesis tests should be made at the beginning of the study. Furthermore, trapping periods need to be synchronous among the populations in order to avoid confounding estimates of animal abundance with temporal shifts in mortality and recruitment.

Unlike closed-population surveys, the catch index (r) in open populations no longer provides a useful index of abundance in tests of hypotheses because in open populations, r is a complex function of not only abundance and capture probabilities but also survival probabilities and recruitment. For example, in a three-period mark–recapture survey of an open population, the expected value of the number of distinct animals caught (r) during the survey is

$$E(r)$$
$$= N_1 p_1 + [N_1(1 - p_1)S_1 + B_1]p_2 + [(N_1(1 - p_1)S_1 + B_1)(1 - p_2)S_2 + B_2]p_3$$

where N_1 = animal abundance at time of first sample
         p_i = probability of capture in period $i (i = 1, \ldots, 3)$
         S_i = probability animal survives from trapping period i to
            $i + 1(i = 1, 2)$
         B_i = number of animals added to the population during interval i
            to $i + 1$ that are still active at time $i + 1(i = 1, 2)$

Expected value of r is further complicated if age-specific mortality and natality are operating in the population. Nichols and Pollock (1983) found similar results for the expected value of the catch index based on the calendar-of-captures method. In either case, valid tests of effects on abundance of open population cannot be performed using these catch indices.

The only suitable catch index in the study of open populations is the number of animals caught in a specific sampling period (n_i) where $E(n_i) = N_i p_i$. Homogeneity of the p_i values across populations is necessary for a test of effects on animal abundance to be valid. A test of homogeneity of the p_i values can be constructed using the method of estimating p_i in the Manly–Parr (1968) method (Figure 5.7). Defining

    c_i = number of animals caught in the ith period that are also caught both before and after period i
    z_i = number of animals not caught in ith period but caught both before and after period i

Figure 5.7 Schematic of capture histories used in estimating capture probabilities by the Manly–Parr (1968) method. Horizontal bars indicate periods when animals in counts c_i and z_i were caught.

the probability of capture can be estimated as

$$\hat{p}_i = \frac{c_i}{c_i + z_i}$$

by identifying a subset of individuals known to be alive in period i $(C_i = c_i + z_i)$. A chi-square test of homogeneity of the p_i values can therefore be based on the $2 \times l$ contingency table of the form

Population 1	Population 2		Population l	
c_1	c_2	$\ldots$	c_l	$c.$
z_1	z_2	$\ldots$	z_l	$z.$
C_1	C_2		C_l	$C.$

In general, however, we do not recommend using the n_i values as catch indices in tests of abundance because of the power $(1 - \beta)$ to test homogeneity of the p_i values across populations will likely be low.

The lack of a suitable catch index for open populations limits the design and analysis of tests of effects on open populations to the use of estimates of absolute abundance. The greater sampling variance of most open-population estimates compared to closed-population surveys reemphasizes the importance of sample size calculations. Power calculations presented in Chapters 5 and 6 can be used in conjunction with open-population models as long as an explicit expression for sampling variance [i.e., $\text{Var}(\hat{N} \mid N)$] exists.

Temporal Dimension of Wildlife Experiments
Throughout this chapter, experimental designs considered have measured population responses at only one point in time after application of the treatments. However, population effects can have a long-term influence on population dynamics, and consequently the temporal dimension of field experiments cannot be ignored. The repeated use of the univariate methods presented above is not recommended for analyzing repeated survey data over time because results of such tests of hypotheses are not independent. The lack of independence comes from the temporal correlation in abundance through time, and will exist regardless of whether independent population surveys are conducted in successive sampling periods. This problem of nonindependence can be lessened, but not eliminated, by fewer and more widely spaced surveys.

The timing of the posttreatment surveys should depend on whether treatment effects are anticipated to produce a short-term transient effect or result in new equilibrium levels for population abundance. From a variety of population growth models, the time predicted for a population to return to a new equilibrium is commonly of the order $1/\theta$, where θ is the instantaneous rate of population growth (May 1981, pp. 6–13). In studies of habitat management practices or population regulation, it is the new equilibrium levels that are typically of primary interest to investigators; hence, there may be objective reasons for the selection of a single experimental endpoint.

Alternatively, a response variable may be selected whose measure of treatment effect is unaffected by the timing of the posttreatment survey. Such invariance occurs with the exponential growth rate (i.e., $dN/dt = \theta N$), which predicts density invariance through time when treatments affect the instantaneous growth rate (θ). In more complex growth models, this temporal invariance occurs only after new equilibrium levels have been achieved.

The use of proportional abundance $K = N_C/N_T$ as a response variable at the single time point in a paired design is based on a model that invokes spatial (and hence, density) invariance of such a ratio at the given point in time. Implicit in such a model is either the assumption that this ratio is density-invariant at all times [i.e., $N_C(t)/N_T(t) = K(t)$] or that the dynamics are sufficiently well understood to identify a time point t where such spatial invariance is expected to hold. Since pair-to-pair spatial variance in K contributes to the experimental error variance at a chosen point in time, lack of spatial invariance degrades the merits of $\bar{K}$ as a measure of treatment effect with model misspecification. An ideal response measure would be one that is both spatially and temporally invariant [i.e., $K(t) = K$] and thus permits inferences to treatment effect in both dimensions. These more demanding model assumptions will be used in Chapter 6 on assessment studies.

Recommendations

Simulation results (Skalski 1985a) indicate that test statistics can be developed to analyze capture data in a hypothesis-testing framework using standard statistical distributions (t and Z). This prospect enables researchers to design experiments and analyze capture data with techniques familiar to most investigators. Furthermore, preliminary investigations of likelihood ratio tests and nonparametric procedures (Skalski 1985a, pp. 216–233) suggest that little may be gained from these approaches in the analysis of population data.

Perhaps the most important aspect of these results is the guidance now available for design of manipulative experiments on animal populations.

Power calculations can be used to estimate the probable success of proposed field studies and identify experiments that have little or no chance of succeeding. Beyond simply identifying feasible experiments, optimization procedures described in this chapter can improve efficiency of field activities. Hopefully, the consequence of this methodology will be stronger experimental inferences and a better use of resources devoted to field investigations.

In addition to the quantitative guidance power calculations and optimization procedures provide, practical considerations can suggest some basic characteristics of all well-designed manipulative experiments. Properties an investigator should incorporate into an experiment to assess effects on abundance of animal populations include

1. Balanced designs providing equal replication for all treatments tested
2. Study plots of equal size and same configuration unless density estimates will be used in tests of effects
3. Equal trapping effort (trap density and survey duration) at all study plots to enhance prospects of interpopulation homogeneity of capture probabilities
4. Populations surveyed simultaneously, or at least sequentially in a randomized block design, to enhance the prospect of interpopulation homogeneity
5. Population surveys performed synchronously in multiplot open-population studies to prevent confounding changes in animal abundance with differences in mortality and recruitment processes
6. Equal dispersal of replicate study plots for all treatments to enhance the prospect of homogeneous treatment variances (randomization ensures equal dispersal in expectation)
7. Use of absolute abundance estimators to quantify abundance at study plots and to serve as a fail-safe feature if interpopulation homogeneity is not achieved
8. Use of survey designs for abundance estimation that permit estimation under various model assumptions and allow goodness-of-fit tests
9. Avoiding use of catch indices in open-population investigations
10. Selection of study plots for an experiment in a manner that permits valid inferences to the target population of interest
11. Preliminary survey data to provide initial parameter estimates for reliable projection of the statistical power and performance of an experiment
12. Principles of randomization and replication at the study plot level of experimentation

To date (Table 5.1), few wildlife experiments have incorporated all or most of the practices recommended above. It is important that investigators reexamine these experimental procedures in light of available knowledge presented here and elsewhere (e.g., Cochran and Cox 1957, Cox 1958). Much can be learned from observational studies of nature, but much more can be deduced if ecological insight is coupled with careful experimentation.

6

Environmental Assessment Studies

The need to assess environmental or human-induced effects on wild populations often occurs in circumstances that prevent or limit the use of randomization and replication of treatment conditions. However, inability to perform a manipulative experiment does not diminish the importance of such investigations. Indeed, some of the more important environmental issues of the day such as effects of lake acidification, hazardous waste releases, and construction of energy production facilities cannot typically be tested in an experimental context. To misconstrue such investigations on wild populations as manipulative experiments runs the risk of performing an ineffectual field study with improper and unwarranted inferences. Alternatively, to withhold management decisions based solely on the relative merits of assessment versus experimental results is to be oblivious of the inherent constraints in environmental assessment studies and of the nature of scientific inquiry into effect assessment. In this chapter, we explore the basis for design and analysis of constrained investigations and the nature of statistical inference from such tests of effects.

R. A. Fisher, one of the early investigators in experimental design and applied statistics, provided an example (Fisher 1947; Bliss 1967, pp. 1–12) of an assessment study (Figure 1.2) we will be using to contrast with a manipulative experiment. The objective of Fisher's study was to determine whether a particular woman could distinguish between two different tea preparations as she claimed. One preparation consisted of pouring the milk into the teacup before pouring the tea; the other preparation was the reverse order. Repeated samples of these two tea preparations were tasted by the woman during blind trials in an effort to test the woman's taste discrimination. Because the study had but a single experimental unit (the woman who made the initial claim), the investigation was an assessment of effects of tea preparation on taste preference of the woman and not an experiment on the effects of different tea preparations on taste preferences of women. In the test, order in which the woman sampled the various cups of tea with different tea preparations was randomized. It was this process of randomization of treatment conditions that permitted the significance level of the test to be calculated and formed the basis for statistical inference.

Fisher obviously had an opportunity to bring additional people into the study, but the choice to limit the investigation to a particular woman was deliberate. On the basis of his objectives, an assessment study was the best design for the investigation. A manipulative experiment would have been more time-consuming and resulting statistical inferences, inappropriate with the stated objective. He was interested in testing the claims of this particular woman and designed the investigation appropriately. As such, an assessment study was not a second-rate alternative to a manipulative experiment, but rather the first and best choice for the investigation.

A parallel can be seen between Fisher's taste test and certain wildlife investigations. During a visit to a hazardous chemical waste site, there was an observed need to keep waterfowl from using a large holding pond that was toxic to wildlife. To determine the effectiveness of carbide cannons and other avoidance tactics to scare waterfowl away from the pond, an assessment study could have been performed. During a series of daily trials, avoidance devices could have been randomly activated or left silent, and numbers of birds that lighted on the pond enumerated. A test of the effects of carbide cannons on waterfowl usage of the pond would then have been a matter of analyzing count data from a two-treatment, completely randomized design (use of "blocking on time" may also have been appropriate). As in Fisher's study, the statistical inference would have been to a particular experimental unit selected for its immediate importance. However, in both assessment studies, the proper design and analysis of the investigation may have suggested a hypothesis of wider interest that could be approached experimentally.

Another important distinction that must be considered is the difference between a treatment hypothesis and an impact hypothesis. By the process of randomizing treatment conditions to experimental units, test conditions are free from confounding influences of locality and time. Thus, the experimental process itself ensures straightforward inferences in manipulative experiments and assessment studies (Figure 1.2). However, when test conditions are systematically or haphazardly administered to study plots, study designs cannot guarantee test results free from confounding influences. This is the situation in impact studies and impact assessments (Figure 1.2). *Model-dependent assumptions and analyses are necessary to separate treatment effects from the potentially confounding influences of time and locality in tests of impact.* The heavy reliance on model-dependent assertions makes the analysis of environmental impact studies particularly challenging. In impact assessments, this challenge becomes particularly acute when testing hypotheses in the absence of true replication or randomization (Figure 1.2).

Approaches to design and analysis of impact and impact assessment studies on wild populations are presented below. Although these designs are intended to have general applicability, the primary purpose is to demonstrate the operating characteristics of such investigations. Typically, there is an array of alternative study designs; the preferred approach is one that incorporates the most information about the status and trends of wild populations under control conditions.

An Impact Study Design

Scenario

Often, wildlife biologists must attempt to evaluate the effects of a potential environmental disturbance under circumstances where they have little or no control of test conditions or its locations. Despite a lack of randomization of test conditions, if the test conditions are replicated across the landscape, a sound inference to effects may be possible. Previous examples we have cited include power transmission lines, road salting, and pesticide spraying. In each case, test conditions of interest are applied repeatedly to the landscape, but in localities of particular advantage or need of the application. As such, the test conditions may be confounded with unique features associated with site selection criteria.

One approach to testing for impact is to use what may be called an "odd's ratio design." The name comes from the similarity of this test of impact to the odd's ratio (Fleiss 1981, pp. 36–37) used in analyzing a 2×2 contingency table. The experimental design consists of l control–treatment pairs of populations similar to the paired design of Chapter 5. However, at each control (C)–treatment (T) pair, animal abundance is estimated twice at each plot instead of just once. During the pretreatment period (0), population abundance is estimated at each control and treatment site. The treatment of interest is then applied to the treatment-designated population of each pair (in a nonrandomized manner) and all populations are then resurveyed during the posttreatment period (1). Thus, use of the odd's ratio design is predicated on prior knowledge of the intended schedule and localities of treatment application.

From each pair of sites, four observations on population abundance, N_{C0}, N_{C1}, N_{T0}, and N_{T1} are used in an odd's ratio design. A test of treatment effects is then based on the average odd's ratio among replicate pairs, where

$$D_i = \frac{N_{C0_i} N_{T1_i}}{N_{C1_i} N_{T0_i}} \tag{6.1}$$

is the odd's ratio for the ith pair, $i = 1, \ldots, l$. By taking a ln-transformation of Eq. (6.1),

$$\ln D_i = \ln N_{C0_i} + \ln N_{T1_i} - \ln N_{C1_i} - \ln N_{T0_i} \tag{6.2}$$

an odd's ratio is seen to be the contrast for a treatment-by-period interaction in a 2×2 factorial treatment design.

The null hypothesis tested by the odd's ratio design is that the mean odd's ratio is equal to the value 1

$$H_0: \mu_D = 1 \tag{6.3}$$

against the alternative hypothesis of

$$H_a: \mu_D \neq 1.$$

Under ln-transformation, the null hypothesis can be written as

$$H_0: \mu_{\ln D} = 0 \tag{6.4}$$

against the alternative hypothesis of

$$H_a: \mu_{\ln D} \neq 0.$$

An appropriate one-tailed hypothesis may be tested if the anticipated direction of the treatment effect is known in advance.

As in all field studies, utility of the inferences from an odd's ratio design will depend on the manner in which sites are selected. Lack of randomization does not abolish the need for proper site selection. Designated treatment sites should be a random or a stratified random sample of sites destined to receive the treatment in order to permit inferences back to this target population. Unlike the paired design of Chapter 5, the control and treatment sites are not from the same target population. The target population in an odd's ratio design is the collection of all sites that receive the treatment. The need for the distinction between designs is that because of lack of randomization in odd's ratio designs, treatment-designated sites may possess characteristics that by their very nature make them candidates for the treatment and different from controls (e.g., in road salting programs, sites on north-facing slopes may be favored).

In contrast, selection of control sites can be more deliberate and less probabilistic. The sole purpose of control sites in an odd's ratio is to serve as a baseline on which change in treatment abundance $(\hat{N}_{T1}/\hat{N}_{T0})$ is compared. This interpretation can be seen by rewriting Eq. (6.2) as

$$\ln D_i = \ln\left(\frac{N_{T1_i}}{N_{T0_i}}\right) - \ln\left(\frac{N_{C1_i}}{N_{C0_i}}\right).$$

For each treatment site, a control area should be selected whose response to seasonal changes from time t to $t + 1$ is anticipated to be similar to that of the treatment site. The appropriate length of time between the two survey periods will depend on a number of factors, including the nature of the effect investigated, objective of the test (e.g., short-term vs. long-term effects), and the synchrony of animal populations. As the time between t and $t + 1$ increases, synchrony between control and treatment populations is likely to decrease, making model-dependent inferences less reliable.

Test Statistics

The model equation for the statistical analysis of an odd's ratio design can be written as

$$\hat{D}_i = \mu\epsilon_i,$$

where $\hat{D}_i$ is the estimated odd's ratio for animal abundance at the ith pair $i = 1, \ldots, l$, μ is the mean odd's ratio for control–treatment pairs, and ϵ_i represents the random-error term $(i = 1, \ldots, l)$. A ln-linear model equation can be written in the form

$$\ln \hat{D}_i = \ln \mu + \ln \epsilon_i.$$

To understand the inferential basis for using an odd's ratio, the nature of D_i needs to be modeled further.

From inspection of Eq. (6.1), D_i is seen to be a ratio of population levels for two different times and places. Let N_{ghi} denote animal abundance at the gth treatment site $(g = C, T)$, during the hth survey period $(h = 0, 1)$, for the ith control–treatment pair $(i = 1, \ldots, l)$. Assuming that spatial and temporal effects have a multiplicative influence on animal abundance, population levels during the surveys can be modeled as

$$N_{ghi} = \mu\tau_g\delta_h\nu_{gi}\epsilon_{ghi}, \tag{6.5}$$

where μ = overall mean abundance
τ_g = effect of the gth treatment, $g = C, T$
δ_h = effect of the hth survey time, $h = 0, 1$
ν_{gi} = effect of the gth test plot in the ith site pair, $i = 1, \ldots, l$
ϵ_{ghi} = random-error term

Substituting response model (6.5) into the expression for the odd's ratio (6.1), it becomes

$$D_i = \frac{(\mu_{T_C} \delta_0 \nu_{Ci} \epsilon_{C0i})(\mu_{T_T} \delta_1 \nu_{Ti} \epsilon_{T1i})}{(\mu_{T_C} \delta_1 \nu_{Ci} \epsilon_{C1i})(\mu_{T_C} \delta_0 \nu_{Ti} \epsilon_{T0i})} = \frac{\tau_T}{\tau_C} \epsilon_i,$$ (6.6)

which in the absence of a treatment effect (i.e., under the null hypothesis $\tau_C = \tau_T$) has an expected value of

$$E(D_i) = 1.$$

Ability to eliminate confounding effects of plot location (ν_{gi}) and time (δ_h) through the calculation of an odd's ratio is why this impact design provides a useful test of effects in the absence of randomization. In so doing, however, the assumption of multiplicative effects of model (6.5) is implied each time an odd's ratio design is analyzed to test the impact hypothesis (6.3). Contrast this observation with the analysis of the paired design of Chapter 5. By randomization, location effects are incorporated into the random error term, and the population response model simplifies to

$$N_{gi} = \mu_{T_g} \epsilon_{gi}.$$

Proportional abundance for a site pair under randomization then becomes

$$K = \frac{N_C}{N_T} = \frac{\mu_{T_C} \epsilon_{Ci}}{\mu_{T_T} \epsilon_{Ti}} = \frac{\tau_C}{\tau_T} \epsilon_i,$$ (6.7)

once again a measure of the relative treatment effect. Although expression (6.7) simplifies as does expression (6.6), it is unnecessary to assume that multiplicative but confounding factors that may mask treatment effects cancel through the use of a data formulation. Consequently, use of randomization provides a robust estimate of treatment effects that can be approached in impact designs only through model-dependent assumptions.

Estimation of the odd's ratio D_i, $i = 1, \ldots, l$ can be based on the use of either absolute abundance estimates or catch indices. When interpopulation heterogeneity in capture probabilities exists, the odd's ratio estimate should be based on absolute abundance estimates where

$$\hat{D}_i = \frac{\hat{N}_{C0i} \hat{N}_{T1i}}{\hat{N}_{C1i} \hat{N}_{T0i}}.$$ (6.8)

When the capture probabilities can be shown to be homogeneous between control and treatment populations within a site pair, the odd's ratio can be estimated more simply as

$$\hat{D}_i = \frac{r_{C0i} r_{T1i}}{r_{C1i} r_{T0i}}.$$ (6.9)

The assumption of interpopulation homogeneity within an odd's ratio can be fulfilled in any one of three ways. Using chi-square tests based on the contingency tables (3.11), (3.15), or (3.20), tests of homogeneity can be performed to determine whether

Case 1: Surveys corresponding to populations N_{C0}, N_{C1}, N_{T0}, and N_{T1} are heterogeneous

Case 2a: Surveys corresponding to populations N_{C0}, N_{C1}, N_{T0}, and N_{T1} are homogeneous

Case 2b: Surveys for populations N_{C0} and N_{C1} are homogeneous and those of N_{T0} and N_{T1} are homogeneous

Case 2c: Surveys for populations N_{C0} and N_{T0} are homogeneous and those of N_{C1} and N_{T1} are homogeneous.

The test for case 2a investigates a more restrictive set of assumptions than either of the pairwise comparisons suggested in the alternative tests of homogeneity (i.e., case 2b and 2c). For valid estimation of the odd's ratio using catch indices, either of the latter two scenarios for case 2 is sufficient. To increase the power of the tests of homogeneity, results of chi-square tests can be pooled (i.e., a sum of chi-square variates is also chi-square-distributed with degrees of freedom equal to sum of degrees of freedom) across similar pairs of sites.

The statistic to test the null hypothesis (6.4) can be written as

$$d = \frac{\frac{1}{l} \sum_{i=1}^{l} \ln \hat{D}_i}{\sqrt{\frac{S_{\ln \hat{D}}^2}{l}}}, \qquad (6.10)$$

where d is approximately t-distributed with $(l - 1)$ degrees of freedom under H_0 using either estimates (6.8) or (6.9) for the odd's ratio.

Example Analysis

Prior to application of a pesticide, five plots destined to be treated were randomly selected from potential spray sites and paired with control plots in close physical proximity. The pretreatment surveys began 10 days prior to the pesticide spraying with the posttreatment surveys commencing 5 days after application. Duration of the surveys was 8 days, four days for initial marking followed by a 4-day recapture period.

Representative capture data for the meadow vole (*Microtus pennsylvanicus*) are given, including both the pre- and posttreatment surveys at control and treatment sites (Table 6.1). Data are arranged in a manner suitable for analysis using Lincoln Index estimates. The posttreatment surveys and capture data were handled independently of pretreatment surveys. Cap-

Table 6.1

Capture Data, Abundance Estimates (3.5), and Estimates of Odd's Ratio (6.8 and 6.9) for *Microtus pennsylvanicus* at Five Control–Treatment Pairs in an Odd's Ratio Design to Assess Effects of a Pesticide Spraying Program on Local Mammal Abundance

Site pair	Control plot Pre-					Control plot Post-					Treatment plot Pre-					Treatment plot Post-					Test of homogeneity		Odd's ratio estimates	
	n_1	n_2	m	r_{C0}	$\hat{N}_{C0}$	n_1	n_2	m	r_{C1}	$\hat{N}_{C1}$	n_1	n_2	m	r_{T0}	$\hat{N}_{T0}$	n_1	n_2	m	r_{T1}	$\hat{N}_{T1}$	(χ_6^2)	α	$\hat{D}_i(6.8)$	$\hat{D}_i(6.9)$
1	34	36	8	62	142.89	30	32	10	52	92.00	37	35	15	57	84.50	9	12	3	18	31.50	4.2281	0.65	0.5790	0.3765
2	45	31	13	63	104.14	31	31	8	54	112.78	30	42	13	59	94.21	19	17	7	29	44.00	8.4007	0.21	0.4313	0.5734
3	22	29	6	45	97.57	23	16	5	34	67.00	36	38	16	58	83.88	16	12	5	23	35.83	8.0641	0.23	0.6221	0.5248
4	21	23	6	38	74.43	23	22	6	39	77.86	19	22	6	35	64.71	13	14	5	22	34.00	0.9201	0.99	0.5023	0.6125
5	29	31	8	52	105.67	21	26	7	40	73.25	44	44	20	68	95.43	15	16	5	26	44.33	4.5927	0.60	0.6702	0.4971

$$\sum \chi_6^2 = 26.2057 \qquad \bar{D} = 0.5610 \qquad \bar{D} = 0.5169$$

ture records during the initial surveys were not used in subsequent abundance estimates during the posttreatment period.

The objective of this study was to determine the short-term effects of the pesticide on small-mammal abundance. To allow for the possible mitigating influence of recolonization, the test of impact was based on separate mark–recapture surveys. Had the objective of the investigation been to test for effects on survival rates, the marked populations from the pretreatment surveys would have been the primary focus, and alternative field sampling and analyses would have been used (Smith and Skalski 1989).

Tests of homogeneity indicate (Table 6.1) that the four-population surveys contributing to estimation of an odd's ratio exhibited similar capture probabilities at each site pair ($\alpha \geq 0.21$). The pooled test of interpopulation homogeneity was also nonsignificant $[P(\chi^2_{30} \geq 26.2057) = 0.66]$, permitting the estimation of the odd's ratio using catch indices according to Eq. (6.9). The nature of the pesticide treatment would suggest a reduction in abundance if an effect occurred. Consequently, the null hypothesis of no impact, $H_0: \mu_D \geq 1$, was rejected $[P(t_4 \leq -8.0248) = .0007]$ in favor of the one-sided alternative

$$H_a: \mu_D < 1.$$

The estimate of μ_D across the five-site pairs (i.e., $\hat{\mu}_D = 0.5169$) suggests roughly a 48% reduction in abundance at treatment sites, the result of pesticide application. Note that all five estimates (Table 6.1) of the odd's ratio, $\hat{D}_i$, $i = 1, \ldots, 5$, are < 1. A sign test (Hollander and Wolfe 1973, pp. 39–45) would similarly reject the null hypothesis in favor of the one-sided alternative at an α-level of $1/2^5 = 0.03125$.

Designing an Odd's Ratio Study

By using the power of the test of null hypothesis (6.4) as an objective function, level of field replication (l) and capture probabilities (p) that result in an effective field design can be determined. The power of an odd's ratio design to detect an impact from a nonrandomized treatment effect will depend on the following: (1) test statistic employed; (2) α-level; (3) level of field replication (l); (4) capture probabilities (p_1 and p_2); (5) magnitude of treatment difference $|\mu_D - 1|$; (6) magnitude of spatial variances, σ^2_{C0}, σ^2_{T0}, σ^2_{C1}, and σ^2_{T1}; and (7) covariance structure between populations, $\text{Cov}(N_{C0}, N_{T0})$, $\text{Cov}(N_{C1}, N_{T1})$, $\text{Cov}(N_{C0}, N_{C1})$, $\text{Cov}(N_{T0}, N_{T1})$, $\text{Cov}(N_{T0}, N_{C1})$, and $\text{Cov}(N_{C0}, N_{T1})$.

The need for estimates of covariances in abundance through time and across landscape is an aspect of study design not encountered in the manipulative experiments of Chapter 5. The covariances of importance in an odd's ratio are defined as follows:

Cov(N_{C0}, N_{T0}) = spatial covariance in abundance between control and treatment plots during pretreatment period

Cov(N_{C1}, N_{T1}) = spatial covariance in abundance between control and treatment plots during posttreatment period

Cov(N_{C0}, N_{C1}) = temporal covariance in abundance between pre- and posttreatment periods at control plots

Cov(N_{T0}, N_{T1}) = temporal covariance in abundance between pre- and posttreatment periods at treatment plots

Cov(N_{C0}, N_{T1}) = covaraince between abundance at control plots during pretreatment and treatment plots during posttreatment

Cov(N_{T0}, N_{C1}) = covariance between abundance at treatment plots during pretreatment and control plots during posttreatment

These covariances can alternatively be expressed as functions of variances and correlations in abundance where, in general, for any two random variables, X and Y:

$$\text{Cov}(X, Y) = \rho_{X,Y}\sqrt{\sigma_X^2\sigma_Y^2}.$$

In this way, the magnitude of a covariance is expressible in terms of population variances and a correlation coefficient that is bounded between $-1 \leq \rho \leq 1$. In population studies, the correlations in abundance can generally be expected to be positive ($0 \leq \rho \leq 1$).

Power Calculations for Case 1 When capture probabilities are expected to be heterogeneous between the four Lincoln Index surveys comprising an odd's ratio, the D_i, $i = 1, \ldots, l$ must be estimated using absolute abundance estimates. The power of the test of (6.4) under case 1 can be estimated from the noncentrality parameter

$$\phi = \frac{\sqrt{l}\,|\ln \mu_D'|}{\sqrt{2\sigma_{EX}^2}}, \tag{6.11}$$

where

$$\mu_D' = \frac{\mu_{C0}(1 - e^{-\mu_{C0}p_1p_2})\mu_{T1}(1 - e^{-\mu_{T1}p_1p_2})}{\mu_{C1}(1 - e^{-\mu_{C1}p_1p_2})\mu_{T0}(1 - e^{-\mu_{T0}p_1p_2})} \tag{6.12}$$

and

$$\sigma_{EX}^2 = \frac{q_1q_2}{p_1p_2}\left\{\left(\frac{1}{\mu_{C0}} + \frac{\sigma_{C0}^2}{\mu_{C0}^3}\right) + \left(\frac{1}{\mu_{T1}} + \frac{\sigma_{T1}^2}{\mu_{T1}^3}\right) + \left(\frac{1}{\mu_{C1}} + \frac{\sigma_{C1}^2}{\mu_{C1}^3}\right) + \left(\frac{1}{\mu_{T0}} + \frac{\sigma_{T0}^2}{\mu_{T0}^3}\right)\right\}$$
$$+ \left\{\sigma_{C0}^2\left(\frac{1}{\mu_{C0}} + \frac{\sigma_{C0}^2}{\mu_{C0}^3}\right)^2 + \sigma_{T1}^2\left(\frac{1}{\mu_{T1}} + \frac{\sigma_{T1}^2}{\mu_{T1}^3}\right)^2\right.$$

$$+ \sigma_{C1}^2 \left(\frac{1}{\mu_{C1}} + \frac{\sigma_{C1}^2}{\mu_{C1}^3} \right)^2 + \sigma_{T0}^2 \left(\frac{1}{\mu_{T0}} + \frac{\sigma_{T0}^2}{\mu_{T0}^3} \right)^2$$

$$+ 2\, \text{Cov}(N_{C0}, N_{T1}) \left(\frac{1}{\mu_{C0}} + \frac{\sigma_{C0}^2}{\mu_{C0}^3} \right) \left(\frac{1}{\mu_{T1}} + \frac{\sigma_{T1}^2}{\mu_{T1}^3} \right)$$

$$+ 2\, \text{Cov}(N_{C1}, N_{T0}) \left(\frac{1}{\mu_{C1}} + \frac{\sigma_{C1}^2}{\mu_{C1}^3} \right) \left(\frac{1}{\mu_{T0}} + \frac{\sigma_{T0}^2}{\mu_{T0}^3} \right)$$

$$- 2\, \text{Cov}(N_{C0}, N_{C1}) \left(\frac{1}{\mu_{C0}} + \frac{\sigma_{C0}^2}{\mu_{C0}^3} \right) \left(\frac{1}{\mu_{C1}} + \frac{\sigma_{C1}^2}{\mu_{C1}^3} \right)$$

$$- 2\, \text{Cov}(N_{C0}, N_{T0}) \left(\frac{1}{\mu_{C0}} + \frac{\sigma_{C0}^2}{\mu_{C0}^3} \right) \left(\frac{1}{\mu_{T0}} + \frac{\sigma_{T0}^2}{\mu_{T0}^3} \right)$$

$$- 2\, \text{Cov}(N_{C1}, N_{T1}) \left(\frac{1}{\mu_{C1}} + \frac{\sigma_{C1}^2}{\mu_{C1}^3} \right) \left(\frac{1}{\mu_{T1}} + \frac{\sigma_{T1}^2}{\mu_{T1}^3} \right)$$

$$- 2\, \text{Cov}(N_{T0}, N_{T1}) \left(\frac{1}{\mu_{T0}} + \frac{\sigma_{T0}^2}{\mu_{T0}^3} \right) \left(\frac{1}{\mu_{T1}} + \frac{\sigma_{T1}^2}{\mu_{T1}^3} \right) \Bigg\} \tag{6.13}$$

and where ϕ corresponds to a noncentral F-distribution with 1 and $(l - 1)$ degrees of freedom. Tables for the noncentral F-distribution (Appendixes 2 and 3) can be used to evaluate Eq. (6.11) as a function of field replication (l) and capture probabilities. Equations (6.12) and (6.13) are written with average capture probabilities $\bar{p}_1$ and $\bar{p}_2$ and project the power of the test based on estimates of absolute abundance.

Power calculations based on Eq. (6.11) are adjusted for potential bias in abundance estimation by Lincoln Indices with the use of (6.12) instead of μ_D. Despite this adjustment, when values of $\mu_N p_1 p_2 \leq 4$ (Robson and Regier 1964), experimental error (σ_{EX}^2) will be overestimated by Eq. (6.13) and the power of the test $(1 - \beta)$ will be underestimated.

Power Calculations for Case 2a   Power calculations for an odd's ratio design also can be computed assuming the catch indices (r) provide valid estimates of D_i, $i = 1, \ldots, l$. Under such circumstances, power of the test of the null hypothesis (6.3) can be estimated from the noncentrality parameter

$$\phi = \frac{\sqrt{l}\, |\ln \mu_D|}{\sqrt{2\sigma_{EX}^2}}, \tag{6.14}$$

where

$$\sigma_{EX}^2 = \frac{q_1 q_2}{1 - q_1 q_2} \Bigg\{ \left(\frac{1}{\mu_{C0}} + \frac{\sigma_{C0}^2}{\mu_{C0}^3} \right) + \left(\frac{1}{\mu_{T1}} + \frac{\sigma_{T1}^2}{\mu_{T1}^3} \right) + \left(\frac{1}{\mu_{C1}} + \frac{\sigma_{C1}^2}{\mu_{C1}^3} \right)$$

$$+ \left(\frac{1}{\mu_{T0}} + \frac{\sigma_{T0}^2}{\mu_{T0}^3} \right) \Bigg\} + \Bigg\{ \sigma_{C0}^2 \left(\frac{1}{\mu_{C0}} + \frac{\sigma_{C0}^2}{\mu_{C0}^3} \right)^2 + \sigma_{T1}^2 \left(\frac{1}{\mu_{T1}} + \frac{\sigma_{T1}^2}{\mu_{T1}^3} \right)^2$$

$$+ \sigma_{C1}^2 \left(\frac{1}{\mu_{C1}} + \frac{\sigma_{C1}^2}{\mu_{C1}^3} \right)^2 + \sigma_{T0}^2 \left(\frac{1}{\mu_{T0}} + \frac{\sigma_{T0}^2}{\mu_{T0}^3} \right)^2$$

$$+ 2\text{Cov}(N_{C0}, N_{T1}) \left(\frac{1}{\mu_{C0}} + \frac{\sigma_{C0}^2}{\mu_{C0}^3} \right) \left(\frac{1}{\mu_{T1}} + \frac{\sigma_{T1}^2}{\mu_{T1}^3} \right)$$

$$+ 2\ \text{Cov}(N_{C1}, N_{T0}) \left(\frac{1}{\mu_{C1}} + \frac{\sigma_{C1}^2}{\mu_{C1}^3} \right) \left(\frac{1}{\mu_{T0}} + \frac{\sigma_{T0}^2}{\mu_{T0}^3} \right)$$

$$- 2\ \text{Cov}(N_{C0}, N_{C1}) \left(\frac{1}{\mu_{C0}} + \frac{\sigma_{C0}^2}{\mu_{C0}^3} \right) \left(\frac{1}{\mu_{C1}} + \frac{\sigma_{C1}^2}{\mu_{C1}^3} \right)$$

$$- 2\ \text{Cov}(N_{C0}, N_{T0}) \left(\frac{1}{\mu_{C0}} + \frac{\sigma_{C0}^2}{\mu_{C0}^3} \right) \left(\frac{1}{\mu_{T0}} + \frac{\sigma_{T0}^2}{\mu_{T0}^3} \right)$$

$$- 2\ \text{Cov}(N_{C1}, N_{T1}) \left(\frac{1}{\mu_{C1}} + \frac{\sigma_{C1}^2}{\mu_{C1}^3} \right) \left(\frac{1}{\mu_{T1}} + \frac{\sigma_{T1}^2}{\mu_{T1}^3} \right)$$

$$- 2\ \text{Cov}(N_{T0}, N_{T1}) \left(\frac{1}{\mu_{T0}} + \frac{\sigma_{T0}^2}{\mu_{T0}^3} \right) \left(\frac{1}{\mu_{T1}} + \frac{\sigma_{T1}^2}{\mu_{T1}^3} \right) \Bigg\} \tag{6.15}$$

and

$$\mu_D = \frac{\mu_{C0}\mu_{T1}}{\mu_{C1}\mu_{T0}} \tag{6.16}$$

associated with a noncentral F-distribution with 1 and $(l - 1)$ degrees of freedom. As in Eq. (6.13), a common estimate of capture probability across site pairs was used in calculating the projected experimental error (6.15).

EXAMPLE On the basis of preliminary survey estimates, historical data or best guesses, and tables in Appendixes 2 and 3, noncentrality parameters (6.11) and (6.14) can be evaluated to project the power of the test with various combinations of l and p. Through repeated power calculations, a combination or combinations of l or p can be found that provide sufficient sampling effort to detect an ecologically important change in animal abundance. For example, consider the case of designing an investigation to detect effects of slash burning on deer mice (*Peromyscus maniculatus*) abundance utilizing scheduled burns to test for an impact. The power of a test of impact based on a field design with six replicate site pairs, a per-period capture probability of $p = 0.6$ during the Lincoln Index surveys, and population parameters as given in Table 6.2, can be computed from Eq. (6.11) where

$$\mu_D' = \frac{50(1 - e^{-50(0.6)^2})25(1 - e^{-25(0.6)^2})}{50(1 - e^{-50(0.6)^2})50(1 - e^{-50(0.6)^2})} = 0.49994$$

and

$$\sigma_{EX}^2 = \left\{ \frac{(0.4)^2}{(0.6)^2} \left[\left(\frac{1}{50} + \frac{130}{50^3} \right) + \left(\frac{1}{25} + \frac{60}{25^3} \right) + \left(\frac{1}{50} + \frac{130}{50^3} \right) + \left(\frac{1}{50} + \frac{130}{50^3} \right) \right] \right\}$$

$$+ \left\{ 130 \left(\frac{1}{50} + \frac{130}{50^3} \right)^2 + 130 \left(\frac{1}{50} + \frac{130}{50^3} \right)^2 \right.$$

$$+ 130\left(\frac{1}{50} + \frac{130}{50^3}\right)^2 + 60\left(\frac{1}{25} + \frac{60}{25^3}\right)^2$$

$$+ 2\left[18\left(\frac{1}{50} + \frac{130}{50^3}\right)\left(\frac{1}{25} + \frac{60}{25^3}\right)\right] + 2\left[26\left(\frac{1}{50} + \frac{130}{50^3}\right)\left(\frac{1}{50} + \frac{130}{50^3}\right)\right]$$

$$- 2\left[78\left(\frac{1}{50} + \frac{130}{50^3}\right)\left(\frac{1}{50} + \frac{130}{50^3}\right)\right] - 2\left[52\left(\frac{1}{50} + \frac{130}{50^3}\right)\left(\frac{1}{50} + \frac{130}{50^3}\right)\right]$$

$$- 2\left[18\left(\frac{1}{50} + \frac{130}{50^3}\right)\left(\frac{1}{25} + \frac{60}{25^3}\right)\right] - 2\left[35\left(\frac{1}{50} + \frac{130}{50^3}\right)\left(\frac{1}{25} + \frac{60}{25^3}\right)\right]\right\}$$

$$= 0.17886$$

leading to

$$\phi_{1,5} = \frac{\sqrt{6}\,|\ln(0.49994)|}{\sqrt{2(0.17886)}} = 2.8393.$$

Reading from tables of the noncentral F-distribution (Appendix 2), projected power is in excess of $1 - \beta > 0.95$ for a two-tailed test at $\alpha = 0.10$ (or one-tailed test at $\alpha = 0.05$).

Generalizations of Power Calculations   For survey models other than the Lincoln Index, the power of an odd's ratio design based on absolute abundance estimates can be computed from the noncentrality parameter (6.14) with the error variance estimated by

$$\sigma_{EX}^2 = \left\{\overline{\text{Var}(\hat{N}_{C0_i})}\left(\frac{1}{\mu_{C0}} + \frac{\sigma_{C0}^2}{\mu_{C0}^3}\right)^2 + \overline{\text{Var}(\hat{N}_{T1_i})}\left(\frac{1}{\mu_{T1}} + \frac{\sigma_{T1}^2}{\mu_{T1}^3}\right)^2\right.$$

$$\left. + \overline{\text{Var}(\hat{N}_{C1_i})}\left(\frac{1}{\mu_{C1}} + \frac{\sigma_{C1}^2}{\mu_{C1}^3}\right)^2 + \overline{\text{Var}(\hat{N}_{T0_i})}\left(\frac{1}{\mu_{T0}} + \frac{\sigma_{T0}^2}{\mu_{T0}^3}\right)^2\right\}$$

$$+ \{\text{2nd term of Eq. (6.13)}\}.$$

This algorithm ignores any small sample bias associated with absolute abundance surveys and assumes $\hat{N}$ approximately ln-normally distributed.

Table 6.2

Population Parameters Used in Designing an Odd's Ratio Study on the Effects of Slash Burning on Deer Mice (*Peromyscus maniculatus*) Abundance

Mean abundance (μ_N)	Population	Variance–covariance matrix			
		N_{C0}	N_{C1}	N_{T0}	N_{T1}
50	N_{C0}	130	78 ($\rho = 0.6$)	52 ($\rho = 0.4$)	18 ($\rho = 0.2$)
50	N_{C1}		130	26 ($\rho = 0.2$)	18 ($\rho = 0.2$)
50	N_{T0}			130	35 ($\rho = 0.4$)
25	N_{T1}				60

When odd's ratios can be estimated using catch indices (i.e., r values), the error variance to use in conjunction with noncentrality parameter (6.14) can be expressed as

$$\sigma_{EX}^2 = \left\{ \frac{1-P}{P} \left[\left(\frac{1}{\mu_{C0}} + \frac{\sigma_{C0}^2}{\mu_{C0}^3} \right) + \left(\frac{1}{\mu_{T1}} + \frac{\sigma_{T1}^2}{\mu_{T1}^3} \right) \right. \right.$$
$$\left. \left. + \left(\frac{1}{\mu_{C1}} + \frac{\sigma_{C1}^2}{\mu_{C1}^3} \right) + \left(\frac{1}{\mu_{T0}} + \frac{\sigma_{T0}^2}{\mu_{T0}^3} \right) \right] \right\}$$
$$+ \{\text{2nd term of Eq. (6.15)}\}$$

where P is the overall probability of capture during the course of the survey. For both absolute abundance and capture indices, the degrees of freedom associated with the noncentral F-distribution is 1 and $l - 1$.

Odd's Ratio Design in Manipulative Experiments

Without the benefits of randomization, there is little alternative to the use of the odd's ratio design if valid inferences to treatment effects are desired. However, in the presence of randomization, odd's ratio design remains a valid study approach and an alternative to the paired design of Chapter 5. To appreciate this conclusion, the response model (6.5) must be used once again. By the process of randomization, location effects (ν_{gi}) are assimilated into the random-error term and Eq. (6.5) becomes

$$N_{ghi} = \mu\tau_g\delta_h\epsilon_{ghi}. \qquad (6.17)$$

Evaluation of an odd's ratio then simplifies to

$$D_i = \frac{(\mu\tau_C\delta_0\epsilon_{C0i})(\mu\tau_T\delta_1\epsilon_{T1i})}{(\mu\tau_C\delta_1\epsilon_{C1i})(\mu\tau_C\delta_0\epsilon_{T0i})}$$
$$= \frac{\tau_T}{\tau_C}\epsilon_i.$$

No longer is it necessary with response model (6.17) to assume that location effects are canceled through their multiplicative influence on abundance. An uncluttered estimate of relative treatment effects is obtained by simply assuming that effects of time cancel. This latter assumption is often made plausible by scheduling the pre- and posttreatment surveys close in time.

However, when randomization is feasible, both the paired design (Chapter 5) and an odd's ratio design may be used to test for treatment effects. Indeed, the posttreatment surveys among the control and treatment plots of an odd's ratio design (i.e., N_{C1}, N_{T1}) constitute what has been referred to as a "paired design." Consequently, paired and odd's ratio designs are competing approaches to manipulative experiments that employ blocking.

Power calculations for the slash burning example above can be repeated for a paired design. The corresponding paired design with six replicate site pairs and an average capture probability of $p = 0.6$ has the noncentrality parameter (5.31) where

$$\mu_k' = \frac{50}{25}\left(\frac{1 - e^{-50(0.6)^2}}{1 - e^{-25(0.6)^2}}\right) = 2.00025$$

and

$$\sigma_{\text{EX}}^2 = \frac{(0.4)^2}{(0.6)^2}\left[\left(\frac{1}{50} + \frac{130}{50^3}\right) + \left(\frac{1}{25} + \frac{60}{25^3}\right)\right]$$
$$+ \left[130\left(\frac{1}{50} + \frac{130}{50^3}\right)^2 + 60\left(\frac{1}{25} + \frac{60}{25^3}\right)^2 - 2(18)\left(\frac{1}{50} + \frac{130}{50^3}\right)\left(\frac{1}{25} + \frac{60}{25^3}\right)\right]$$
$$= 0.16849$$

leading to

$$\phi_{1,5} = \frac{\sqrt{6}\,|\ln(2.00025)|}{\sqrt{2(0.16849)}} = 2.9253.$$

Hence, the paired design with half the number of population surveys has greater power than the corresponding odd's ratio design (i.e., $\phi_{1,5} = 2.9253 > 2.8393$). Although surprising, this result is not uncommon.

To appreciate how a field study with greater effort can have poorer performance than a comparable but simpler experiment, the structure of the error variances must be investigated. According to the data in Table 6.2, and a per-period capture probability of $p = 0.6$, the odd's ratio design has a projected experimental error (6.13) of $\sigma_{\text{EX}}^2 = 0.17886$ versus a smaller variance (5.33) of $\sigma_{\text{EX}}^2 = 0.16849$ for the paired design. Inspection of experimental error variances (5.33) and (6.13) shows that each error variance is composed of two components, sampling error and a measure of the natural heterogeneity in K_i or D_i between site pairs. In the slash burning example, variance (5.33) has the constituents $\sigma_{\text{EX}}^2 = 0.02883 + 0.13966$, whereas in the odd's ratio design, $\sigma_{\text{EX}}^2 = 0.04754 + 0.13132$, in terms of sampling error and heterogeneity, respectively. Thus, for the odd's ratio design, additional information in the pretreatment observations did not exceed the increased sampling error introduced by the added surveys. This behavior of an odd's ratio design results from the artifact of using mark–recapture survey methods.

If population abundance was enumerated rather than estimated, there would be no sampling error [i.e., $\text{Var}(\hat{N}\,|\,N) = 0$] to contribute to σ_{EX}^2, and the resulting error variance would typically decrease with incorporation of the pretreatment data in an odd's ratio design. This often would be the case in plant ecology and benthic studies where entire study plots are surveyed or harvested. However, with the typical large sampling errors of wildlife surveys, the anticipated performance of an odd's ratio design is not easily predicted without the aid of power calculations (6.11) or (6.14).

Not all comparisons of the relative efficiency of odd's ratio versus paired designs are as counterintuitive as the preceding example. The variance–covariance structure of an anticipated study can be such that major gains in power can be achieved using odd's ratio designs in wildlife investigations. Careful evaluation of the economics and statistical behavior of tests of effects is necessary in making the proper design choice. Yet another example will be used to illustrate this point (Figure 6.1). Power curves based on the parameters in Table 6.3 indicate a range of values for l and p wherein the paired design for equal replication (l) is more powerful than the odd's ratio design and yet other values for which the odd's ratio design is superior (Figure 6.1). The relative power of the odd's ratio design increases over a paired design with equal replication as sampling error decreases (i.e., $p \rightarrow 1$). This behavior is again a consequence of the sampling error in surveying animal abundance.

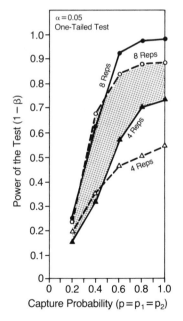

Figure 6.1 Comparison of power curves ($\alpha = 0.05$ one-tailed or $\alpha = 0.10$ two-tailed) for paired (dashed lines) and odd's ratio (solid lines) designs based on population parameters (Table 6.3). Shaded area indicates the loss of power in an odd's ratio design ($l = 4$) relative to a paired design ($l = 8$) with the same survey costs.

An even more appropriate comparison might be to compare the power of an odd's ratio design with l-pairs and a paired design with $2l$-pairs (Figure 6.2). In such circumstances, each design uses the same number of $4l$ surveys in test of effects. When survey costs are the principal expense in a field study, and site preparation and treatment application are inexpensive, such comparisons provide a first-order approximation of the relative cost-efficiency of the two study approaches. In Figure 6.1, a comparison of the odd's ratio design with $l = 4$ and that of the paired design with $l = 8$ shows the paired design to have a 0.15–0.3 increase in power for the same per-period capture probability. Consequently, the paired design in this case (Table 6.3) is a better study approach despite initial impressions based on power calculations alone. When the costs of plot establishment and application of treatments are appreciable, design optimization as illustrated in Chapter 5 should be used to determine the most cost-effective design between paired and odd's ratio options.

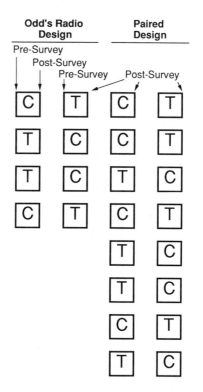

Figure 6.2 Graphical representation of the temporal and spatial layout of an odd's ratio design with $l = 4$ site pairs and a paired design with $l = 8$ site pairs illustrating the equal number (16) of mark–recapture surveys.

Table 6.3

Anticipated Population Parameters Used in Designing Either a Paired or Odd's Ratio Study
in the Presence of Randomization of Treatments to Study Sites

Mean abundance (μ_N)	Population	\multicolumn{4}{c}{Variance–covariance matrix}
		N_{C0}	N_{C1}	N_{T0}	N_{T1}
100	N_{C0}	580	464 ($\rho = 0.8$)	348 ($\rho = 0.6$)	197 ($\rho = 0.4$)
100	N_{C1}		580	232 ($\rho = 0.4$)	296 ($\rho = 0.6$)
100	N_{T0}			580	296 ($\rho = 0.6$)
75	N_{T1}				420

Cost of Not Randomizing Treatments

One significant cost of not randomizing treatments to plots is that inferences
to impacts must be model-dependent. As such, the appropriateness of the
response model (6.5) over the range of conditions observed is vital to valid
statistical inferences. When randomization is used, such potentially confounding factors as plot location are experimentally eliminated, and most
model-dependent assertions are no longer necessary [i.e., response model
(6.17)]. Hence, conclusions from randomized designs will generally be
more reliable and less controversial.

The second cost of not randomizing treatments to plots is a potential
loss of efficiency when paired designs must be abandoned in favor of the
odd's ratio design. The difference in power of 0.15–0.3 between an odd's
ratio design with $l = 4$ and a paired design with $l = 8$ (Figure 6.1) may be
regarded as an approximate measure of this loss of efficiency. However,
this loss must be tempered with the realities of field research. An impact
study may have the opportunity to take advantage of test conditions that
would otherwise go uninvestigated or be too cost-prohibitive to study experimentally. There are occasions in field research where the only options are
to conduct a nonrandomized study or not perform the investigation at all.
Under such conditions, the loss of statistical power relative to an ideal has
little meaning. Odd's ratio design provides a basis for the inferences to effects in the absence of randomization.

Design Optimization

Field studies using odd's ratio designs can be optimized for choice of plot
size (A), number of plots (l), trapping effort (f), and survey duration (d) by
the same graphical process considered for completely randomized design in
Chapter 5. The form of the cost function will be similar, and the appropriate noncentrality parameter is that given in Eq. (6.11) or (6.14). For case 1,
the optimal trapping effort and level of replication (l) can be found from er-

ror variance (6.13) and catch–effort model $p = 1 - e^{-cf/A}$ by graphing solutions to the equation

$$f = \frac{A}{c} \ln\left[1 + \frac{1}{\sqrt{\dfrac{l(\ln \mu_D)^2}{2G\phi^2} - \dfrac{H}{G}}}\right] \qquad (6.18)$$

where G = first bracketed term ({ }) in Eq. (6.13) associated with the sampling error

H = second bracketed term ({ }) in Eq. (6.13) associated with heterogeneity in animal abundance

c = catch coefficient in the catch–effort relationship
$p = 1 - e^{-cf/A}$

$\mu_D = \dfrac{\mu_{C0}\mu_{T1}}{\mu_{C1}\mu_{T0}}$

ϕ = value of the noncentrality parameter for an F-distribution with l and $(l - 1)$ degrees of freedom at specified values of α and β

Using the appropriate values of ϕ (Appendixes 2 and 3) in Eq. (6.18), odd's ratio designs may be readily optimized. The similarity in the form of ϕ for paired and odd's ratio designs allows the use of Eq. (6.18) with either design by an appropriate specification of terms G and H.

In the situation of homogeneous capture probabilities (i.e., case 2a), the plot of the power curve as a function of trap effort is based on equation

$$f = \frac{A}{2c} \ln\left[1 + \frac{1}{\left[\dfrac{l(\ln \mu_D)^2}{2G\phi^2} - \dfrac{H}{G}\right]}\right]$$

where G and H are as previously defined. Optimal allocation is found from the point where the cost function and power curve share a common tangent.

An Impact Assessment Design

Scenario
The need to make a site-specific assessment of a potential impact on faunal abundance introduces yet another constraint on the use of classical experimental designs in wildlife investigations (Figure 1.2). To compensate for a lack of replication and randomization, the study design needs to incorporate as much prior knowledge concerning the nature of the potential impact as

possible. Green (1979, pp. 71–73) provides a useful decision rule for selecting the best strategy for an impact assessment based on extent of prior knowledge concerning the "whens" and "wheres" of the impact. The most favorable situation is what Green (1979, pp. 68–71) calls his "optimal design." This optimal strategy requires baseline (i.e., preoperational) data to be collected along with the establishment of spatial control sites.

To detect an environmental impact, the study design must be able to differentiate between natural fluctuations in faunal abundance and changes as a result of impact. The detection of differences in faunal abundance between control and treatment stations or between preoperational and operational phases of monitoring is insufficient evidence alone for assigning causation to population change. Both spatial and temporal controls are usually necessary to establish a cause–effect relationship in impact assessment. Green's "optimal design" provides the best opportunity for establishing such a cause–effect relationship. Consequently, we have chosen this strategy as the first of the impact assessment designs to be presented. In the next section (i.e., "An Accident Assessment Design"), impact assessment design in the absence of baseline data will be presented.

A design that incorporates both spatial and temporal components in an impact assessment is the control–treatment paired (CTP) design of Skalski and McKenzie (1982). The basic advantage of CTP designs is their use of spatial controls to account for temporal effects on dynamics systems. Treatment sites are located near the potential source of impact, whereas control sites are positioned outside the zone of potential effects. As in classical experimental designs, control areas measure the effects of ambient conditions, and treatment areas measure the same environmental conditions plus effects of the hypothesized additional stimulus, the impact.

To account for temporal fluctuations in animal abundance, control–treatment pairs are established during preoperational phase of site development, and these station pairs are sampled through the operational period. The preoperational sampling serves to evaluate success of the pairing scheme and establishes the relationship of faunal abundance between control and treatment stations prior to impact. The observed relationship of abundance between control and treatment sites is later compared to that observed during operational monitoring. An impact is defined using this scenario as a statistically significant change in the proportional abundance ($K = N_C/N_T$) at control–treatment stations between preoperational and operational phases (Chapman 1951, Eberhardt 1976).

Success of CTP designs depends on faunal populations at control and treatment sites "tracking" each other, that is, maintaining a constant proportionality. Subtle microhabitat differences prevent perfect pairing of sites in practice, and monitoring concepts must be used that permit the realization there is no such thing as widely separated, yet truly "replicate," study sites. The CTP design does not require exact pairing; populations simply need to "track" each other. Such synchrony among populations is common, because local populations tend to respond to similar climatic and environmental conditions.

Test Statistics

A CTP investigation is conceptualized as a completely randomized experimental design with a factorial treatment design. The model equation for analysis can be written as

$$\hat{K}_{ijkh} = \mu \gamma_i \theta_k \rho_h \epsilon_{ijkh},$$ (6.19)

where $\hat{K}_{ijkh}$ = estimated ratio of faunal abundance at control and treatment plots (N_C/N_T) for the ith pair $(i = 1, \ldots, l)$ during the kth season $(k = 1, \ldots, s)$ of the jth year $(j = 1, \ldots, t)$ and the hth operational phase $(h = 1, 2)$

μ = mean of ratios for control–treatment pairs

γ_i = location effect for the ith pair $(i = 1, \ldots, l)$

θ_k = seasonal effect on proportional abundance $(k = 1, \ldots, s)$

ρ_h = effect of operational status on proportional abundance $(h = 1, 2)$

ϵ_{ijkh} = random error term $(j = 1, \ldots, t)$

Using a logarithmic transformation of proportional abundance, a linearized model results:

$$\ln \hat{K}_{ijkh} = \ln \mu + \ln \gamma_i + \ln \theta_k + \ln \rho_h + \ln \epsilon_{ijkh},$$

permitting the use of analysis of variance (ANOVA) techniques. This model uses yearly observations of a control–treatment pair as the replication to make site-specific inferences of impact (Skalski and McKenzie 1982). The statistical test of impact is based on a test for significant main effects of operational phase on proportional abundance. As such, the null hypothesis of no impact can be written as

$$H_0: \rho_1 = \rho_2$$ (6.20)

against the alternative

$$H_a: \rho_1 \neq \rho_2.$$

In the analysis of CTP -design data, individual estimates of proportional abundance $(\hat{K})$ may be computed based on catch indices (4.7) or absolute abundance estimates (5.29), depending upon presence of interpopulation homogeneity. The two scenarios considered in the analysis of a CTP design are as follows:

Case 1: Surveys corresponding to populations N_{Ckh} and N_{Tkh} are heterogeneous.

Case 2: Surveys corresponding to populations N_{Ckh} and N_{Tkh} are homogeneous.

Evaluation of pairwise homogeneity should be performed on each site pair separately for each survey period. The most appropriate estimate for proportional abundance should be selected separately for each estimate of K used in CTP designs. With estimates of proportional abundance computed, ANOVA techniques are used to analyze the K values and test the null hypothesis (6.20). The general form of the ANOVA can be depicted by a degree-of-freedom table (Table 6.4). The F-statistic used in the test is then

$$
F = \frac{(X_{\ldots 1} - X_{\ldots 2})^2 / 2slt}{\left[\sum_{h=1}^{2} \sum_{k=1}^{s} \sum_{j=1}^{t} \sum_{i=1}^{l} (X_{ijkh} - \bar{X}_{i \cdot kn})^2 \right] \Big/ 2sl(t - 1)} = \frac{\text{MSP}}{\text{MSE}}
\tag{6.21}
$$

where $X_{ijkh} = \ln \hat{K}_{ijkh} = $ log ratio of proportional abundance at control and treatment plots (i.e., N_C/N_T) for the ith pair ($i = 1, \ldots, l$) during the kth season ($k = 1, \ldots, s$) of the jth year ($j = 1, \ldots, t$) and the hth operational phase ($h = 1, 2$) and is distributed as an F-statistic with 1 and $2sl(t - 1)$ degrees of freedom under the null hypothesis (Skalski 1984). Simulation studies (Skalski 1985a) show statistic (6.21) to be F-distributed using both indices and absolute abundance estimates.

Table 6.4

Degree-of-Freedom Table Associated with the ANOVA (Based on a Balanced Design) for a CTP Design and Statistical Test of Impact

Source	df	SS	MS	F
Total	$2slt$	SST		
Mean	1	SSM		
Total corrected	$(2slt - 1)$	SSTOT		
Main Effects	$(l + s - 1)$			
Location (L)	$(l - 1)$	SSL	MSL	
Season (S)	$(s - 1)$	SSS	MSS	
Phase (P)	1	SSP	MSP	F = MSP/MSE
Interactions	$(2sl - l - s)$			
L × S	$(l - 1)(s - 1)$			
L × P	$(l - 1)$			
S × P	$(s - 1)$			
L × S × P	$(l - 1)(s - 1)$			
Error	$2sl(t - 1)$	SSE	MSE	

Lacking true replication (i.e., only one facility), model-dependent assumptions are necessary in order to test for effects [i.e., test hypothesis (6.20)] using CTP design. As a substitute for true replication, repeated yearly observations of proportional abundance ($K = N_C/N_T$) at a site pair are used to measure temporal variation of this metric and provide a site-specific assessment of effects. For these yearly observations of proportional abundance to serve the role of replicates, the observations must be identically distributed about a common mean. For this to occur, the annual effects on animal abundance must be eliminated from the measured responses so that resulting observations are replicate measurements of the same population response. Using the definition of proportional abundance (K) and the population response model (6.5), temporal effects on animal counts are seen to be eliminated by use of this metric where

$$K_{hi} = \frac{N_{Chi}}{N_{Thi}} = \frac{\mu \tau_C \delta_h \nu_{Ci} \epsilon_{Chi}}{\mu \tau_T \delta_h \nu_{Ti} \epsilon_{Thi}}$$

$$= \frac{\tau_C \nu_{Ci}}{\tau_T \nu_{Ti}} \epsilon_{hi} . \qquad (6.22)$$

Repeated estimates of K through time at a site pair thus provide independent estimates of this metric, whose expected value will change only through a change in relationship between τ_C and τ_T with plant operations. During preoperational monitoring, $\tau_C = \tau_T$ and remains as such during operational monitoring under null hypothesis (6.20). If an impact does occur, $\tau_C \neq \tau_T$, a change in the average proportional abundance ($\bar{K}_{1i} \neq \bar{K}_{2i}$) between preoperational and operational monitoring periods at a site pair is anticipated.

From relationship (6.22), two model-dependent assumptions are seen to exist in the analysis of CTP designs. First, temporal effects on animal abundance must be multiplicative (or additive if differences are analyzed, $N_{Chi} - N_{Thi}$) so that they are eliminated by the calculation of proportional abundance ($\hat{K}$). Second, although δ_h may differ between pairs, the magnitude of the temporal effects is assumed to be the same for control and treatment stations within a site pair. Tests of additivity mentioned earlier (Chapter 1) are useful here in testing these assumptions concerning the nature of temporal effects and the choice of proper data formulation.

Consequently, model-dependent assumptions are once again necessary to test impact hypotheses. Need to assume multiplicative effects and synchronous populations may be considered the price for making a site-specific assessment in the absence of randomization and true replication. Eberhardt (1976) has called the CTP design a "pseudoexperiment" because of this false replication. However, unlike many of the designs suggested by Green (1979), the test of impact is not based on sampling error alone. The experimental error variance includes both sampling error and the variance of the metric (i.e., $\hat{K}_i$) through time. One effect of imperfect synchrony between populations is that the magnitude of the error variance becomes unduly inflated. Inflation of the variance estimate provides a degree of robustness for the test of impact under violations in assumptions of population synchrony and strict multiplicativeness. The trade-off under such circumstances is a loss of power in tests of impact.

Example Analysis

A test of impact for near-field effects of ultra-high-voltage transmission lines on surrounding wildlife populations is a simple illustration of the application of the CTP design (Figure 6.3). Consider a case where six control–treatment pairs are established along the length of a prototype transmission line, with three pairs established on either side. Treatment plots would be established near the base of the line but outside the zone of physical disturbance of the right-of-way. Control plots would be located in line with test plots and perpendicular to the corridor (Figure 6.3). Test plots are spaced such that mark–recapture surveys would not interfere with neighboring sites. Control plots would be located far enough away from the power line to be outside the zone of potential impact, yet close enough to ensure that all populations responded similarly to climatic events. Assume three years of baseline sampling are conducted at this prototype transmission line following construction and before the line is energized.

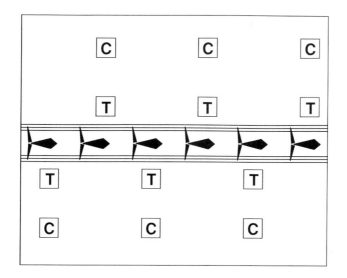

Figure 6.3 Spatial layout of a CTP design to assess impacts along the corridor of an ultra-high-voltage transmission line (not drawn to scale).

Table 6.5

Lincoln Index Estimates ($\hat{N}$) of *Microtus montanus* Abundance (Standard Deviations in Parentheses), and Proportional Abundance ($\hat{K}$) for Six Control–Treatment Pairs of a CTP Design Sampled over a 6-year Period, 3 Years' Preoperational and 3 Years' Operational

Year	Season	Phase*	Pair 1 $\hat{N}_C$	$\hat{N}_T$	$\hat{K}$	Pair 2 $\hat{N}_C$	$\hat{N}_T$	$\hat{K}$	Pair 3 $\hat{N}_C$	$\hat{N}_T$	$\hat{K}$	Pair 4 $\hat{N}_C$	$\hat{N}_T$	$\hat{K}$	Pair 5 $\hat{N}_C$	$\hat{N}_T$	$\hat{K}$	Pair 6 $\hat{N}_C$	$\hat{N}_T$	$\hat{K}$
3	Spring	P	25.00 (6.84)	17.00 (4.24)	1.4706	23.00 (13.42)	31.50 (9.37)	0.7302	12.33 (3.33)	17.67 (5.58)	0.6981	25.00 (4.42)	21.67 (3.45)	1.1538	21.75 (3.08)	15.50 (3.40)	1.4032	14.00 (7.75)	11.00 (4.00)	1.2727
3	Fall	P	129.00 (35.14)	85.40 (17.14)	1.5105	95.25 (13.72)	80.40 (11.59)	1.1847	95.67 (21.23)	67.00 (23.03)	1.4279	137.00 (40.94)	112.14 (29.88)	1.2217	96.50 (23.16)	113.36 (22.76)	0.8512	59.38 (12.79)	37.25 (6.91)	1.5940
2	Spring	P	26.00 (6.00)	14.75 (3.44)	1.7627	38.00 (13.96)	22.40 (5.00)	1.6964	13.00 (2.16)	39.00 (14.49)	0.3333	19.17 (3.46)	19.00 (7.75)	1.0088	13.00 (6.48)	29.00 (10.25)	0.4483	14.00 (5.48)	15.20 (2.94)	0.9211
2	Fall	P	61.33 (16.16)	76.33 (12.98)	0.8035	128.60 (42.49)	69.00 (15.28)	1.8638	145.67 (62.83)	98.89 (18.35)	1.4730	91.44 (20.04)	76.78 (16.21)	1.1910	73.67 (12.38)	45.20 (7.45)	1.6298	33.00 (7.31)	57.50 (19.20)	0.5739
1	Spring	P	26.00 (11.22)	29.33 (6.36)	0.8864	87.00 (42.99)	38.20 (10.29)	2.2775	15.00 (4.47)	40.25 (12.60)	0.3727	17.33 (2.95)	54.00 (25.69)	0.3210	29.00 (10.25)	26.50 (7.60)	1.0943	13.00 (2.90)	14.00 (3.00)	0.9286
1	Fall	P	107.50 (26.33)	95.00 (17.19)	1.1316	94.45 (18.16)	110.86 (29.65)	0.8520	75.36 (13.67)	113.40 (36.90)	0.6646	135.00 (40.40)	108.20 (22.54)	1.2477	107.15 (18.88)	123.00 (36.45)	0.8712	36.33 (8.43)	47.57 (10.62)	0.7638
1	Spring	O	24.71 (4.29)	9.50 (0.87)	2.6015	19.00 (7.75)	41.00 (25.10)	0.4634	14.00 (3.00)	5.00 (1.41)	2.8000	19.00 (4.90)	5.67 (1.05)	3.3529	20.67 (3.80)	15.67 (4.41)	1.3191	15.00 (4.47)	7.00 (3.46)	2.1429
1	Fall	O	45.44 (8.19)	38.00 (18.83)	1.1959	92.50 (16.64)	59.00 (10.70)	1.5678	59.50 (12.83)	107.33 (44.57)	0.5543	74.60 (14.50)	49.67 (12.52)	1.5020	64.57 (9.01)	60.75 (13.03)	1.0629	42.33 (10.06)	24.67 (8.27)	1.7162
2	Spring	O	12.50 (2.60)	5.00 (0.00)	2.5000	17.00 (4.24)	18.25 (4.50)	0.9315	16.50 (3.96)	30.50 (13.56)	0.5410	17.00 (6.93)	14.40 (2.48)	1.1805	30.50 (13.56)	23.50 (10.10)	1.2979	5.25 (0.56)	13.67 (1.87)	0.3841
2	Fall	O	47.89 (8.81)	37.86 (7.90)	1.2650	84.43 (21.53)	31.00 (4.73)	2.7235	39.71 (8.35)	47.86 (10.73)	0.8299	63.00 (18.76)	51.50 (16.36)	1.2233	44.00 (8.37)	45.75 (9.05)	0.9617	26.00 (4.81)	44.00 (20.49)	0.5909
3	Spring	O	35.00 (10.73)	11.00 (4.00)	3.1818	15.71 (1.89)	9.00 (2.23)	1.7460	18.80 (3.98)	9.00 (3.16)	2.0889	32.00 (11.49)	19.00 (10.95)	1.6842	13.40 (2.40)	17.00 (3.46)	0.7882	20.00 (8.37)	9.00 (2.24)	2.2222
3	Fall	O	65.50 (17.57)	64.71 (15.62)	1.0121	85.67 (18.54)	105.67 (44.32)	0.8107	191.00 (25.86)	44.00 (8.66)	4.3409	91.00 (28.77)	41.22 (7.18)	2.2075	74.43 (16.53)	52.20 (14.95)	1.4258	55.00 (21.49)	16.60 (3.25)	3.3133

* P = preoperational; O = operational.

Single mark–recapture data for the Montane vole (*Microtus montanus*), representing 6 years of study, are illustrated (Table 6.5). Data represent semiannual surveys in spring and fall during 3 years of preoperational and 3 years of operational monitoring. As such, each population was surveyed 12 times during the course of the impact assessment. Capture data correspond to a 4-day marking period followed by a 4-day recapture period for each survey. Survey schedules would be simultaneous at site pairs to enhance the prospects that proportional abundance ($\hat{K}_i$) might be estimated more precisely using catch indices.

Using absolute abundance estimates of the Montane vole (Table 6.5), proportional abundance estimates were calculated (5.29) at each site pair and for each survey period. Plots of survey data (Figure 6.4*a*) indicated that

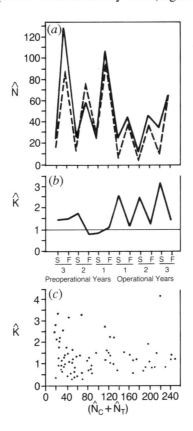

Figure 6.4 Plots of abundance estimates (*a*) for a control and treatment pair of a CTP design (site pair 1, Table 6.5), and the resulting estimates of proportional abundance (*b*) against survey period. Plot of $\hat{K}$ versus $\hat{N}_C + \hat{N}_T$ (all site pairs, Table 6.5) shows no significant ($a > 0.1$) correlation ($r = 0.17$) and suggesting additivity on the log scale (*c*).

control and treatment areas tracked each other quite well (average sample correlation $\bar{r} = 0.36$ over season and operational phase). An additional plot of $\hat{K}$ versus $\hat{N}_C + \hat{N}_T$ (Figure 6.4c) indicates that seasonal effects may have been approximately multiplicative $[P(|r_{70}| \geq 0.017) > 0.1]$ and an analysis of the impact data using ln-differences was appropriate. Further, plots of data (Figures 6.4a, 6.5a) indicate that although seasonal effects are quite evident in the abundance estimates $(\hat{N})$, no such trend exists in the estimates of proportional abundance (Figures 6.4b, 6.5b). As such, assumptions of independence and additivity on the logarithmic scale appear reasonable with the vole data, permitting an analysis using test statistic (6.21).

The F-statistic for the vole data was found to be significant $[P(F_{1,48} > 6.44) = 0.0145]$ and the null hypothesis (6.20) of no impact was consequently rejected. A less formal test based on numbers of estimates of proportional abundance above and below the value of $K = 1$ (Figure 6.5b) between operational phases in a 2×2 contingency table indicated

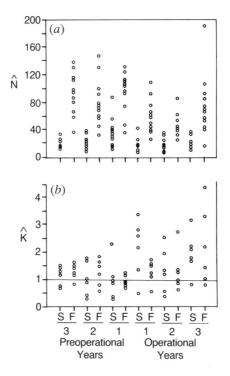

Figure 6.5 Plots of estimates of animal abundance (*a*) and proportional abundance (*b*) against survey period from a CTP design (Table 6.5) to assess the effects of a transmission corridor on local vole abundance. Plots show the elimination of seasonal trends in abundance data $(\hat{N})$ by calculation of $\hat{K}$.

similar results $[P(\chi_1^2 \geq 3.003) = 0.0831]$. The magnitude of the effect of the impact was estimated to be

$$\frac{\hat{K} \cdots 1}{\hat{K} \cdots 2} - 1 = \frac{1.1010}{1.6536} - 1 = -0.3342$$

or, in other words, an estimated 33% reduction in local vole abundance as a result of the transmission corridor.

Designing Control–Treatment Pair Studies

The power of the CTP design to detect an impact will depend on the following: (1) test statistic employed, (2) α-level, (3) number of years of sampling per operational phase (t), (4) extent of seasonal sampling (s), (5) extent of spatial sampling (l), (6) capture probabilities (p_1 and p_2), (7) magnitude of the impact (V), (8) magnitude of the temporal variances in abundance σ_C^2 and σ_T^2, and (9) temporal covariance [$\text{Cov}(N_C, N_T)$] between control and treatment sites through time. It is important to note that temporal variance in abundance rather than the spatial variance influences the power of the test. The CTP design is the only field design presented in this book where such a relationship exists.

Preliminary sample surveys will rarely be conducted long enough (which would be ≥ 2 years) to collect sufficient data to estimate temporal variances. Instead, historical data on population dynamics of species will typically be necessary to provide such estimates. A preliminary survey in conjuction with a CTP design then serves to provide cost data and to model catch–effort relationships important in design optimization. The shift from using spatial to temporal variances in CTP designs also serves to emphasize the importance of designing a preliminary survey consistent with objectives of the consummate field study.

Using tables for the noncentral F-distribution (Appendixes 2 and 3), the power of a proposed impact design can be calculated using the noncentrality parameter corresponding to a noncentral F-distribution with 1 and $2ls(t - 1)$ degrees of freedom:

$$\phi = \frac{\sqrt{slt} \, |\ln(1 + V)|}{2\sqrt{\sigma_{\text{EX}}^2}} \tag{6.23}$$

where l = number of control–treatment pairs
s = number of seasons sampled per year
t = number of years monitored per phase
V = fractional change of treatment populations during the operational phase as a result of impact (e.g., -0.25 corresponds to a 25% reduction in abundance)
σ_{EX}^2 = expected experiment error

As in the previous field designs discussed, the structure of σ_{EX}^2 will change depending on whether measurements of absolute abundance or relative abundance are used in estimating the values.

Power Calculations for Case 1   When proportional abundance is estimated by Eq. (5.29), the experimental error (σ_{EX}^2) in (6.23) is expressible as

$$
\begin{aligned}
\sigma_{EX}^2 = \frac{1}{2s} \sum_{h=1}^{2} \sum_{k=1}^{s} &\left\{ \left[\sigma_{\bar{C}kh}^2 \left(\frac{1}{\mu_{Ckh}} + \frac{\sigma_{\bar{C}kh}^2}{\mu_{\bar{C}kh}^3} \right)^2 + \sigma_{\bar{T}kh}^2 \left(\frac{1}{\mu_{Tkh}} + \frac{\sigma_{\bar{T}kh}^2}{\mu_{\bar{T}kh}^3} \right)^2 \right. \right. \\
&\left. - 2 \, \mathrm{Cov}(N_{Tkh}, N_{Ckh}) \left(\frac{1}{\mu_{Tkh}} + \frac{\sigma_{\bar{T}kh}^2}{\mu_{\bar{T}kh}^3} \right) \left(\frac{1}{\mu_{Ckh}} + \frac{\sigma_{\bar{C}kh}^2}{\mu_{\bar{C}kh}^3} \right) \right] \\
&\left. + \frac{q_1 q_2}{p_1 p_2} \left[\left(\frac{1}{\mu_{Ckh}} + \frac{\sigma_{\bar{C}kh}^2}{\mu_{\bar{C}kh}^3} \right) + \left(\frac{1}{\mu_{Tkh}} + \frac{\sigma_{\bar{T}kh}^2}{\mu_{\bar{T}kh}^3} \right) \right] \right\}
\end{aligned}
\tag{6.24}
$$

where a single common set of capture probabilities (p_1, p_2) are assumed for the Lincoln Indices. The parameters μ_{Tkh}, μ_{Ckh}, $\sigma_{\bar{T}kh}^2$, $\sigma_{\bar{C}kh}^2$, and $\mathrm{Cov}(N_{Ckh}, N_{Tkh})$ in Eq. (6.24) are the temporal moments in faunal abundance specific to a season ($k = 1, \ldots, s$) and an operational monitoring phase ($h = 1, 2$). For instance, if surveys are to be conducted both spring and fall, separate estimates of these five parameters would be necessary for each season and monitoring phase. Hence, a total of 20 ($= 5 \times 2 \times 2$) parameter estimates would be necessary for the power calculations.

Power Calculations for Case 2   When capture probabilities are homogeneous between control and treatment sites within a site pair during the survey periods for a Lincoln Index, catch indices (r) can be used to estimate proportional abundance; the error variance in noncentrality parameter (6.23) becomes

$$
\begin{aligned}
\sigma_{EX}^2 = \frac{1}{2s} \sum_{h=1}^{2} \sum_{k=1}^{s} &\left\{ \left[\sigma_{\bar{C}kh}^2 \left(\frac{1}{\mu_{Ckh}} + \frac{\sigma_{\bar{C}kh}^2}{\mu_{\bar{C}kh}^3} \right)^2 + \sigma_{\bar{T}kh}^2 \left(\frac{1}{\mu_{Tkh}} + \frac{\sigma_{\bar{T}kh}^2}{\mu_{\bar{T}kh}^3} \right)^2 \right. \right. \\
&\left. - 2 \, \mathrm{Cov}(N_{Tkh}, N_{Ckh}) \left(\frac{1}{\mu_{Tkh}} + \frac{\sigma_{\bar{T}kh}^2}{\mu_{\bar{T}kh}^3} \right) \left(\frac{1}{\mu_{Ckh}} + \frac{\sigma_{\bar{C}kh}^2}{\mu_{\bar{C}kh}^3} \right) \right] \\
&\left. + \frac{q_1 q_2}{(1 - p_1 p_2)} \left[\left(\frac{1}{\mu_{Ckh}} + \frac{\sigma_{\bar{C}kh}^2}{\mu_{\bar{C}kh}^3} \right) + \left(\frac{1}{\mu_{Tkh}} + \frac{\sigma_{\bar{T}kh}^2}{\mu_{\bar{T}kh}^3} \right) \right] \right\}
\end{aligned}
\tag{6.25}
$$

assuming a common set of capture probabilities (p_1, p_2). As in case 1 above, the population parameters in Eq. (6.25) must be estimated for all seasons, monitoring phases, and treatment categories.

EXAMPLE   Survey results from estimating vole abundance (Table 6.5) can be used to evaluate the statistical power of that CTP field design. To calculate power of the test of impact, population parameters $[\mu_N, \sigma_N^2, \text{Cov}(N_C, N_T)]$ and capture rates ($\bar{p}$) at the control and treatment sites must be estimated (Table 6.6) for the two seasons (spring and fall) and two operational phases (preoperational and operational) of study. These parameter estimates were derived from the sample moments of the abundance estimates using the relationships

$$\mu_N \triangleq \hat{\hat{N}}, \tag{6.26}$$
$$\sigma_N^2 \triangleq S_N^2 - \overline{\text{Vâr}(\hat{N})}, \tag{6.27}$$

$$\bar{p} \triangleq \frac{1}{1 + \sqrt{\dfrac{\overline{\text{Vâr}(\hat{N})}}{\hat{\mu}_N}}}, \tag{6.28}$$

and

$$\text{Cov}(N_C, N_T) \triangleq \text{cov}(\hat{N}_C, \hat{N}_T) \tag{6.29}$$

for the eight population categories [treatment designations (2) × seasons (2) × operational phase (2)]. These eight categories can be functionally reduced to six by noting that the preoperational and operational status of a facility should have no effect on control sites. In (6.27), s_N^2 was the temporal variance in $\hat{N}$ pooled across sites (Snedecor and Cochran 1980, p. 91).

Using the parameter estimates associated with the vole study (Table 6.6), the noncentrality parameter associated with a field design having six control–treatment pairs, 3 years of biannual surveys per monitoring phase and an average per-period capture probability, for example, of $\bar{p} = 0.3$ is

$$\phi_{1,48} = \frac{\sqrt{2 \cdot 6 \cdot 3}\,|\ln(1 - 0.33)|}{2\sqrt{0.4772}} = 1.7392,$$

Table 6.6

Estimated Population Parameters and Capture Rates Derived from the Abundance Estimates ($\hat{N}$) and Their Variances $[\text{Vâr}(\hat{N}\,|\,N)]$ of Animal Surveys (Table 6.5) for a CTP Design

Monitoring phase	Season	Control sites			Treatment sites			
		$\hat{\mu}_N$	$\hat{\sigma}_N^2$	$\hat{p}$	$\hat{\mu}_N$	$\hat{\sigma}_N^2$	$\hat{p}$	$\hat{\rho}_{N_C, N_T}$
Preoperational	Spring	21.6	56.8	0.320	25.3	28.8	0.346	0.149*
Operational	Spring	21.6	56.8	0.320	14.6	26.9	0.324	0.767
Preoperational	Fall	81.9	101.5	0.227	83.5	100.5	0.298	0.534
Operational	Fall	81.9	101.5	0.227	51.2	195.0	0.280	−0.246
		$\bar{\bar{p}} = 0.274$			$\bar{\bar{p}} = 0.312$			$\bar{\bar{\rho}} = 0.360*$

*Correlation coefficients were estimated by pooling individual estimates across site pairs or sampling periods (Snedecor and Cochran 1980, pp. 187–188).

where

$$
\begin{aligned}
\sigma_{\text{EX}}^2 = \frac{1}{2 \cdot 2} \Bigg\{ &\Bigg[56.8\left(\frac{1}{21.6} + \frac{56.8}{(21.6)^3}\right)^2 + 28.8\left(\frac{1}{25.3} + \frac{28.8}{(25.3)^3}\right)^2 \\
&- 2(0.149)\sqrt{(56.8)(28.8)}\left(\frac{1}{21.6} + \frac{56.8}{(21.6)^3}\right)\left(\frac{1}{25.3} + \frac{28.8}{(25.3)^3}\right) \\
&+ \frac{0.7^2}{0.3^2}\left(\left(\frac{1}{21.6} + \frac{56.8}{(21.6)^3}\right) + \left(\frac{1}{25.3} + \frac{28.8}{(25.3)^3}\right)\right) \Bigg] \\
&+ \Bigg[101.5\left(\frac{1}{81.9} + \frac{101.5}{(81.9)^3}\right)^2 + 100.5\left(\frac{1}{83.5} + \frac{100.5}{(83.5)^3}\right)^2 \\
&- 2(0.534)\sqrt{(101.5)(100.5)}\left(\frac{1}{81.9} + \frac{101.5}{(81.9)^3}\right)\left(\frac{1}{83.5} + \frac{100.5}{(83.5)^3}\right) \\
&+ \frac{0.7^2}{0.3^2}\left(\left(\frac{1}{81.9} + \frac{101.5}{(81.9)^3}\right) + \left(\frac{1}{83.5} + \frac{100.5}{(83.5)^3}\right)\right) \Bigg] \\
&+ \Bigg[56.8\left(\frac{1}{21.6} + \frac{56.8}{(21.6)^3}\right)^2 + 26.9\left(\frac{1}{14.6} + \frac{26.9}{(14.6)^3}\right)^2 \\
&- 2(0.767)\sqrt{(56.8)(26.9)}\left(\frac{1}{21.6} + \frac{56.8}{(21.6)^3}\right)\left(\frac{1}{14.6} + \frac{26.9}{(14.6)^3}\right) \\
&+ \frac{0.7^2}{0.3^2}\left(\left(\frac{1}{21.6} + \frac{56.8}{(21.6)^3}\right) + \left(\frac{1}{14.6} + \frac{26.9}{(14.6)^3}\right)\right) \Bigg] \\
&+ \Bigg[101.5\left(\frac{1}{81.9} + \frac{101.5}{(81.9)^3}\right)^2 + 195.0\left(\frac{1}{51.2} + \frac{195.0}{(51.2)^3}\right)^2 \\
&- 2(-0.246)\sqrt{(101.5)(195.0)}\left(\frac{1}{81.9} + \frac{101.5}{(81.9)^3}\right)\left(\frac{1}{51.2} + \frac{195.0}{(51.2)^3}\right) \\
&+ \frac{0.7^2}{0.3^2}\left(\left(\frac{1}{81.9} + \frac{101.5}{(81.9)^3}\right) + \left(\frac{1}{51.2} + \frac{195.0}{(51.2)^3}\right)\right) \Bigg] \Bigg\} \\
= \; & 0.4772,
\end{aligned}
$$

when the anticipated impact is a 33% reduction in animal abundance. From Appendix 2, the power of the test of impact with the parameter estimates from Table 6.6 is approximately $1 - \beta = 0.79$ at a significance level of $\alpha = 0.05$ (one-tailed test).

Similar calculations can be performed to investigate the power of the test for alternative levels of field sampling (l), duration of the monitoring phase (t), and average capture rates (p). The noncentrality parameter for the CTP design then becomes

$$
\phi_{1, 4l(t-1)} = \frac{\sqrt{2lt}\,|\ln(1 - 0.33)|}{2\sqrt{0.09579 + 0.07005\left(\frac{1}{p} - 1\right)^2}} \tag{6.30}
$$

on the basis of the parameter estimates in Table 6.6. Power curves (Figure 6.6) developed with the use of Eq. (6.30) show the obvious increase in power of the test of impact with increased levels of field sampling (l), capture probabilities (p), and phase duration (t). The vole study would have had a projected power of $1 - \beta = 0.64$ instead of the 0.79 if the study had been conducted for only 2 years per phase rather than $t = 3$. Similarly, by increasing the level

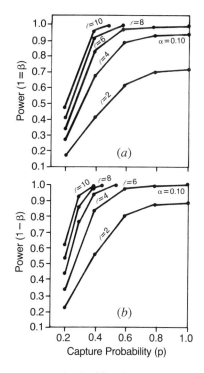

Figure 6.6 Power curves associated with a CTP impact assessment design (two-tailed test $\alpha = 0.10$) based on population parameter in Table 6.6 for different levels of plot replication (l), capture probabilities (p), and study durations of (a) *2 years,* (b) *3 years* of biannual sampling during preoperational and operational monitoring.

of field replication from $l = 6$ to $l = 10$, the power of the test would increase from 0.79 to 0.93. To determine the influence of increasing the number of seasonal sample(s), additional estimates of season-specific population parameters would be needed in Eq. (6.24). In general, the power of the test will increase with additional surveys conducted within a yearly cycle constrained only by the requirement of independence.

An important observation from construction of power curves (Figure 6.6) is that power of impact assessment designs plateau when capture rates approach $p = 0.6$ per period. This phenomenon has been observed previously in other study designs (Figure 5.5). An increase in power is then largely achieved by additional field plots or study duration. Consequently, a few highly characterized study areas may be insufficient to quantify impacts regardless of the amount of capture data collected. Performance of a CTP design depends on the joint nature of the sampling effort of the mark–recapture surveys and the spatial and temporal sampling about the site and the impact episode.

Generalization of Power Calculations   The general form of the error variance for a CTP design with any mark–recapture survey technique can be expressed as

$$\sigma_{EX}^2 = \frac{1}{2s} \sum_{h=1}^{2} \sum_{k=1}^{s} \left\{ \left[\sigma_{Ckh}^2 \left(\frac{1}{\mu_{Ckh}} + \frac{\sigma_{Ckh}^2}{\mu_{Ckh}^3} \right)^2 + \sigma_{Tkh}^2 \left(\frac{1}{\mu_{Tkh}} + \frac{\sigma_{Tkh}^2}{\mu_{Tkh}^3} \right)^2 \right. \right.$$

$$- 2 \, \text{Cov}(N_{Tkh}, N_{Ckh}) \left(\frac{1}{\mu_{Ckh}} + \frac{\sigma_{Ckh}^2}{\mu_{Ckh}^3} \right) \left(\frac{1}{\mu_{Tkh}} + \frac{\sigma_{Tkh}^2}{\mu_{Tkh}^3} \right) \right]$$

$$+ \overline{\text{Var}(\hat{N}_{Ckh})} \left(\frac{1}{\mu_{Ckh}} + \frac{\sigma_{Ckh}^2}{\mu_{Ckh}^3} \right)^2 + \overline{\text{Var}(\hat{N}_{Tkh})} \left(\frac{1}{\mu_{Tkh}} + \frac{\sigma_{Tkh}^2}{\mu_{Tkh}^3} \right)^2 \right\}$$

where $\overline{\text{Var}(\hat{N}_{jkh})}$ is the average sample variance in abundance estimation for the jth treatment $j = (C, T)$ during the kth season ($k = 1, \ldots, s$) of the hth operational phase ($h = 1, 2$). This error variance would be used in conjuction with noncentrality parameter (6.23) associated with a noncentral F-distribution with 1 and $2ls(t - 1)$ degrees of freedom.

In the case where capture probabilities are homogeneous, the error variance in Eq. (6.23) can be expressed as

$$\sigma_{EX}^2 = \frac{1}{2s} \sum_{h=1}^{2} \sum_{k=1}^{s} \left\{ \left[\sigma_{Ckh}^2 \left(\frac{1}{\mu_{Ckh}} + \frac{\sigma_{Ckh}^2}{\mu_{Ckh}^3} \right)^2 + \sigma_{Tkh}^2 \left(\frac{1}{\mu_{Tkh}} + \frac{\sigma_{Tkh}^2}{\mu_{Tkh}^3} \right)^2 \right. \right.$$

$$- 2 \, \text{Cov}(N_{Tkh}, N_{Ckh}) \left(\frac{1}{\mu_{Ckh}} + \frac{\sigma_{Ckh}^2}{\mu_{Ckh}^3} \right) \left(\frac{1}{\mu_{Tkh}} + \frac{\sigma_{Tkh}^2}{\mu_{Tkh}^3} \right) \right]$$

$$+ \frac{1 - P}{P} \left[\left(\frac{1}{\mu_{Ckh}} + \frac{\sigma_{Ckh}^2}{\mu_{Ckh}^3} \right) + \left(\frac{1}{\mu_{Tkh}} + \frac{\sigma_{Tkh}^2}{\mu_{Tkh}^3} \right) \right] \right\}$$

where P is the overall probability of capture during the course of the survey.

Design Optimization

The diminishing return in power to test for impacts with increasing capture rates (Figure 6.6) once again suggests the need to find the most cost-effective study design. In other words, optimization should be used to determine the allocation of field efforts that will maximize the power of the test for fixed costs, or alternatively, the design that will minimize costs for fixed performance. Using the monitoring study of the Montane vole as an example, the study will be reevaluated to determine the effects of optimal allocation on the projected power of the study. The average observed capture probability of $\bar{p} = 0.3$ corresponds to a catch coefficient (2.13) of $c \approx 0.0036$ for a survey period of 4 days with a trap density of 100 traps per 1 ha plot. The objective of the optimization will be to determine the number of paired plots (l) and trapping effort per plot (f) that will maximize the power of the test for the same expenditure (C_0), plot size, and temporal sampling effort (3 years of biannual sampling per phase).

The cost function for the vole study has the same functional form as Eq. (5.22) where

$$C_0 = 100l + 235.11\sqrt{l} + 235lf + 280.35l\sqrt{f} \qquad (6.31)$$

the total variable cost of the investigation is expressed in terms of plot replication (l) and trapping effort (f). The budget for the original study design
($l = 6, f = 100$) then has the projected cost of $169,797 based on cost
function (6.31). The first step in determining the optimal allocation of effort
within the original budget and constraints of cost function (6.31) is to solve
the cost function (6.31) in terms of l and f (dashed line, Figure 6.7). Along
this curve is a combination of l and f that will maximize the power of the
test of impact for the fixed budget of $C_0 = \$169,797$.

To identify the optimal design configuration, solutions to Eq. (6.23)
must be plotted to find the power curve with the greatest value of $1 - \beta$ that
is a tangent to the cost curve. Using the catch–effort relationship (2.13), the
noncentrality parameter (6.23) can be solved in terms of f where

$$f = -\frac{1}{c} \ln\left[\frac{\sqrt{\{slt[\ln(1 + V)]^2/(4\phi^2 B)\} - A/B}}{1 + \sqrt{\{slt[\ln(1 + V)]^2/(4\phi^2 B)\} - A/B}}\right] \qquad (6.32)$$

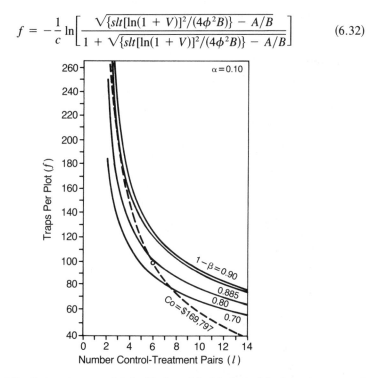

Figure 6.7 Cost contour associated with the original budget of the impact assessment
study [Eq. (6.31), dashed line] and power contour for the test of impact based on the population data in Table 6.6 (solid lines). The optimal design ($l = 3, f = 205$) is identified by the
intersection of the cost contour and the power contour for $1 - \beta = 0.885$. The point on the
cost contour identifies the original design configuration.

and where

$$
A = \frac{1}{2s} \sum_{h=1}^{2} \sum_{k=1}^{s} \left[\sigma_{Ckh}^2 \left(\frac{1}{\mu_{Ckh}} + \frac{\sigma_{Ckh}^2}{\mu_{Ckh}^3} \right)^2 + \sigma_{Tkh}^2 \left(\frac{1}{\mu_{Tkh}} + \frac{\sigma_{Tkh}^2}{\mu_{Tkh}^3} \right)^2 \right.
$$
$$
\left. - 2 \, \mathrm{Cov}(N_{Tkh}, N_{Ckh}) \left(\frac{1}{\mu_{Tkh}} + \frac{\sigma_{Tkh}^2}{\mu_{Tkh}^3} \right) \left(\frac{1}{\mu_{Ckh}} + \frac{\sigma_{Ckh}^2}{\mu_{Ckh}^3} \right) \right]
$$
$$
B = \frac{1}{2s} \sum_{h=1}^{2} \sum_{k=1}^{s} \left[\left(\frac{1}{\mu_{Ckh}} + \frac{\sigma_{Ckh}^2}{\mu_{Ckh}^3} \right) + \left(\frac{1}{\mu_{Tkh}} + \frac{\sigma_{Tkh}^2}{\mu_{Tkh}^3} \right) \right]
$$

based on proportional abundance estimator (5.29) and where ϕ has 1 and $2ls(t-1)$ degrees of freedom. Using Eq. (6.32), the power curve corresponding to $1 - \beta = 0.885$ (Figure 6.7) is found to share a common tangent with the cost contour. Consequently, a maximum power of 0.885 is projected for a study with a budget of $C_0 = \$169,797$ by establishing three control–treatment pairs with each site receiving 205 traps. This optional allocation would have improved the power of the test from 0.77 in original design to 0.885 without additional field costs. As chance would have it, the original design and resulting data detected a statistically significant impact.

By repeated application of the above iterative process with different budgets and power curves, a series of optimal solutions (Figure 6.8) to the design of the monitoring study can be found. As with a previous example of a completely randomized experimental design (Figure 5.6), the optimal design curve reached a threshold once a sufficient trapping effort was achieved. In this vole monitoring study, a trapping effort of 210 traps per 1-ha site corresponds to a per-period capture probability [i.e., Eq. (2.13)] of $p = 0.53$ (78% of a population is expected to be captured during a survey). Once this level of trapping effort and three control–treatment pairs are established, additional resources should be used to increase numbers of paired sites in order to increase the statistical power of the test of impact. Alternatively, the length of the monitoring program could be investigated to determine the most cost-effective duration and level of plot replication to optimize performance of the study. However, construction and production schedules do not permit much latitude in timing and length of an impact assessment. As such, investigators need to determine those design configurations within their prerogative that will maximize the performance of a study.

An Accident Assessment Design

Scenario
Another common need for impact assessment occurs in accident scenarios. Examples of accident assessments include studies of effects of forest fires,

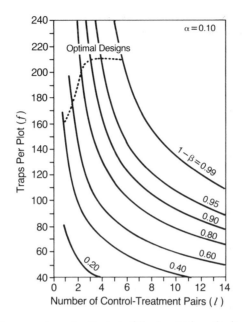

Figure 6.8 Power contours for the test of the impact hypothesis (6.20) based on the population parameters in Table 6.5. Optimal design configurations sketched from identification of individual solutions (dotted line).

natural disasters, and toxic waste spills on wild populations. These scenarios have in common with impact assessment studies the absence of replication and randomization of control and treatment conditions (Figure 1.2). Because of the unexpected nature of an accident, however, the opportunity to establish paired plots or collect baseline data is typically absent. Alternative approaches to impact assessment in accident scenarios are therefore required.

A spatial pattern or gradient in animal abundance centered about the source of the accident would provide *a posteriori* evidence of a biotic response to environmental stress. Typically, numerous sites would have to be surveyed in and about the area of the accident to establish a spatial pattern. The number of sites would typically be too large for use with mark–recapture methods.

Should the effect of the accident be transitory in nature, an alternative assessment strategy based on a treatment-by-time interaction could be used to test the null hypothesis of no impact. The test of impact is based on a comparison of time trends (or profiles) in mean abundance between control and treatment sites over time. Under the null hypothesis of no impact, the time profiles would be parallel for control and treatment sites (Figure 6.9). This parallelism would be a result of regional climatic influences affecting both control and treatment sites in a similar manner over time. Under the alternative hypothesis of impact, the time profiles would not parallel until population levels recover from the impact (Figure 6.9). Because treatments are not randomized to sites, there is no expectation that mean abundance would be equal for control and treatment sites. Rather, relative abundance between treatments may simply reflect inherent differences in carrying capacity between locations within and outside the accident zone.

(a) Under H_o

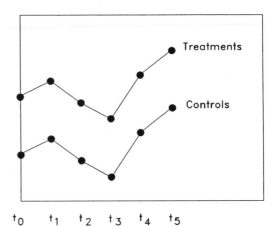

(b) Under H_a

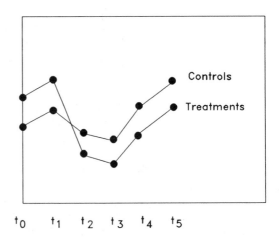

Figure 6.9 Profiles of mean abundance of control and treatment sites over time under (a) the null hypothesis (H_0) of no impact, (b) the alternative hypothesis (H_a) of impact.

A sampling design for testing parallelism in an accident scenario consists of selecting l_C sites outside the influence of the accident zone (i.e., control sites) and l_T sites within the suspected impacted area (i.e., treatment sites). The $l_C + l_T$ sites are then sampled simultaneously during t periods over time. Since the spatial pattern of accident zones is often irregular in shape and may coincide with patterns in the terrain, pairing of control and treatment sites may not be possible. The effects of the accident, as in the case of wildfires, may also minimize the opportunity to match sites based on habitat characteristics after the occurrence. For these reasons, an unpaired design will often be necessary and will be the approach illustrated (Figure 6.10).

The repeated estimation over time of animal abundance at the $l_C + l_T$ sites constitutes repeated measures. A comparison of the time trends of control and treatment sites can be based on multivariate profile analysis (Morrison 1976, pp. 153–160, 205–223). The individual surveys are envisioned as independent and of short duration permitting closed–population estimators [i.e., Lincoln Index, or multiple-mark–recapture models of Otis et al. (1978)]. Nevertheless, the temporal processes of natality and mortality at

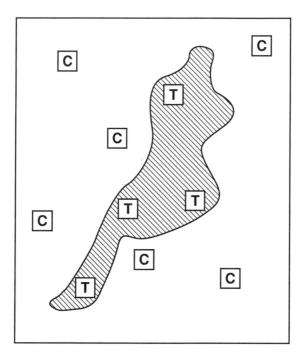

Figure 6.10 Illustration of the spatial array of control (C) and treatment (T) sites about a hypothetical wildfire site.

the study areas will result in abundance (N values) at these sites being correlated through time. This temporal correlation violates the assumption of independence assumed in univariate F- and t-tests and precludes the use of sampling periods as a factor in ANOVA. In the control–treatment paired designs discussed earlier, the use of proportional abundance minimized this temporal correlation. In accident scenarios with unpaired sampling designs, the temporal correlations cannot be ignored, and alternative analyses must be sought.

Test Statistic

In a test of parallelism, population abundance during surveys in an accident assessment can be modeled as a function of spatial, temporal, and treatment effects. The treatment effects in this case are assumed to be transitory, diminishing with time following the accident. Assuming a multiplicative model for population abundance, the response model can be wrriten as

$$\hat{N}_{hij} = \mu \tau_{hi} \delta_i \gamma_{hj} \epsilon_{hij}, \tag{6.33}$$

where μ = overall mean abundance
  τ_{hi} = effect of the hth treatment ($h = C, T$) during the ith survey period ($i = 1, 2, \ldots, t$)
  δ_i = effect of the ith survey period ($i = 1, 2, \ldots, t$)
  γ_{hj} = effect of the jth test plot ($j = 1, \ldots, l_h$) in the hth treatment ($h = C, T$)
  ϵ_{hij} = random-error term

A ln-transformation of the response model (6.33) results in an additive model of the form

$$\ln \hat{N}_{hij} = \ln \mu + \ln \tau_{hi} + \ln \delta_i + \ln \gamma_{hj} + \ln \epsilon_{hij}.$$

The effect of control conditions under this model is assumed to be constant over time $\tau_{C1} = \tau_{C2} = \cdots = \tau_{Ct} = \tau_C$. Under H_0, the treatment effects are assumed of the form $\tau_{T1} = \tau_{T2} = \cdots = \tau_{Tt}$. However, under H_a, treatment effects are assumed to diminish with time such that $\tau_{T1} > \tau_{T2} > \cdots > \tau_{Tt}$.

A test of the null hypothesis of no impact is equivalent to a test of parallelism (or equivalently, a treatment-by-time interaction) in an accident scenario. For example, in an accident assessment with two survey periods, the test of impact is based on the linear contrast

$$(\overline{\ln N_{C1}} - \overline{\ln N_{C2}}) - (\overline{\ln N_{T1}} - \overline{\ln N_{T2}}). \tag{6.34}$$

Substituting the response model (6.33) into contrast (6.34) results in the expression

$$\ln \tau_{T2} - \ln \tau_{T1} = \ln\left(\frac{\tau_{T2}}{\tau_{T1}}\right)$$

which under the null hypothesis (H_0: $\tau_{T1} = \tau_{T2}$) has expected value 0. The more transistory the nature of the impact between survey periods ($i = 1, \ldots, t$), the more likely a test of parallelism will detect on impact. Lettenmaier et al. (1978) found that in the case where an intervention (i.e., accident) results in an immediate effect that decays exponentially over time, the power of the assessment largely depends on the number of postaccident surveys and not on preaccident sampling.

Overall Test of Parallelism   For a multiperiod accident assessment, a test of impact may be based on an overall test of parallelism between control and treatment stations over the duration of the study. The null hypothesis of no impact therefore is written as

$$H_0: \begin{bmatrix} \mu_{C1} - \mu_{C2} \\ \mu_{C2} - \mu_{C3} \\ \vdots \\ \mu_{C,t-1} - \mu_{Ct} \end{bmatrix} = \begin{bmatrix} \mu_{T1} - \mu_{T2} \\ \mu_{T2} - \mu_{T3} \\ \vdots \\ \mu_{T,t-1} - \mu_{Tt} \end{bmatrix} \tag{6.35}$$

against

$$H_a: \begin{bmatrix} \mu_{C1} - \mu_{C2} \\ \mu_{C2} - \mu_{C3} \\ \vdots \\ \mu_{C,t-1} - \mu_{Ct} \end{bmatrix} \neq \begin{bmatrix} \mu_{T1} - \mu_{T2} \\ \mu_{T2} - \mu_{T3} \\ \vdots \\ \mu_{\tau,t-1} - \mu_{Tt} \end{bmatrix}$$

The form of null hypothesis (6.35), although widely applicable, does not take into account the belief of diminishing effects (i.e., $\tau_{T1} > \tau_{T2} > \cdots > \tau_{Ti}$) with time as illustrated in Figure 6.9. More specific tests can be proposed under the specific alternative of transient effects.

In the case of multivariate normal data with equal variance–covariance matrices (i.e., Σ) for controls and treatments, Morrison (1976, pp. 153–160) presents a Hotellings T^2-statistic, or an equivalent F-test of null hypothesis (6.35). A ln-transformation of abundance estimates is recommended to achieve approximate normality and additivity of effect and to stabilize treatment variances. In accident scenarios, treatments are not randomized to sites, and because the zone of potential impact is spatially confined, treatment variances may be unequal (Figure 6.10). Should the spatial dispersion of the study areas prove to have an appreciable effect on within-treatment variance–covariances, an alternative approach to the analysis as illustrated below is necessary.

Sequential Test of Parallelism   If a transitory effect is hypothesized, a sequential series of tests may be performed working from period $i = t$ backward in time to period $i = 1$. The first test of impact would be based on the null hypothesis

$$H_0: \mu_{C,t-1} - \mu_{Ct} = \mu_{T,t-1} - \mu_{Tt} \tag{6.36}$$

against

$$H_a: \mu_{C,t-1} - \mu_{Ct} \neq \mu_{T,t-1} - \mu_{Tt}.$$

If the nature of the impact on animal abundance is known, a one-tailed test may be proposed instead of hypothesis (6.36). Should hypothesis (6.36) not be rejected, the next hypothesis in the sequence tested is

$$H_0: 2\mu_{C,t-2} - \mu_{C,t-1} - \mu_{Ct} = 2\mu_{T,t-2} - \mu_{T,t-1} - \mu_{Tt} \tag{6.37}$$

against

$$H_a: 2\mu_{C,t-2} - \mu_{C,t-1} - \mu_{Ct} \neq 2\mu_{T,t-2} - \mu_{T,t-1} - \mu_{Tt}.$$

Further tests in the sequence are represented by the orthogonal contrasts (Kirk 1982, pp. 92–101) presented in Table 6.7. Tests are performed until the null hypothesis of no impact is rejected or no impact is found. The sequence of tests in this manner provides a means to determine both the presence and the duration of an impact. The $t - 1$ sequential tests, based on the orthogonal contrasts, are approximately uncorrelated. To guard against inflating the experimentwise error rate, the testwise error rate for each sequential test should be adjusted by the relationship

$$\alpha_{EX} = 1 - (1 - \alpha_T)^{t-1}$$

Table 6.7

Orthogonal Constrasts for a Five-Period ($t = 5$) Accident Assessment Based on Sequential Tests of Parallelism in Both Reverse and Forward Directions

Sequence	Contrast	Treatment mean									
		μ_{C1}	μ_{C2}	μ_{C3}	μ_{C4}	μ_{C5}	μ_{T1}	μ_{T2}	μ_{T3}	μ_{T4}	μ_{T5}
Reverse	1	0	0	0	1	−1	0	0	0	−1	1
	2	0	0	2	−1	−1	0	0	−2	1	1
	3	0	3	−1	−1	−1	0	−3	1	1	1
	4	4	−1	−1	−1	−1	−4	1	1	1	1
Forward	1	−1	1	0	0	0	1	−1	0	0	0
	2	−1	−1	2	0	0	1	1	−2	0	0
	3	−1	−1	−1	3	0	1	1	1	−3	0
	4	−1	−1	−1	−1	4	1	1	1	1	−4

or equivalently

$$\alpha_T = 1 - \sqrt[t-1]{1 - \alpha_{EX}}$$

where α_{EX} is the experimentwise error rate and α_T is the testwise error rate. For example, if the desired experimentwise error is $\alpha_{EX} = 0.10$ and the accident assessment consists of $t = 5$ sampling periods, the individual tests of impact should be performed at

$$\alpha_T = 1 - \sqrt[4]{1 - 0.10} = 0.026.$$

A benefit of the sequential tests is that the inherently multivariate analysis is converted to univariate methods.

Sequential tests of impact can be performed based on asymptotic t-statistics with $l_C + l_T - 2$ degrees of freedom. As with all other analyses, tests can be based on absolute abundance estimates when interpopulation heterogeneity exists (i.e., case 1) or catch indices when interpopulation homogeneity (i.e., case 2) can be demonstrated. Using contingency table tests of homogeneity, the form of homogeneity can be determined where:

Case 1: All $l_C + l_T$ population surveys are heterogeneous during each of the t survey periods.

Case 2a: During each survey period, all $l_C + l_T$ populations are homogeneous but capture probabilities are heterogeneous between survey periods.

Case 2b: All population surveys through time and among populations are homogeneous.

For convenience, define the transformed variables under case 1 as

$$x_{ij} = \ln \hat{N}_{Cij}$$
$$y_{ij} = \ln \hat{N}_{Tij}$$

or analogously, under cases 2a and 2b as

$$x_{ij} = \ln r_{Cij}$$
$$y_{ij} = \ln r_{Tij}.$$

To illustrate the test statistics, an accident assessment with three survey periods following an accident will be considered. The surveys will be assumed to have produced the mean vector

$$\hat{\mu} = \begin{bmatrix} \bar{x}_1 \\ \bar{x}_2 \\ \bar{x}_3 \\ \bar{y}_1 \\ \bar{y}_2 \\ \bar{y}_3 \end{bmatrix} \tag{6.38}$$

and the corresponding estimated variance–covariance matrix of the sample means

$$
\hat{\Sigma} = \begin{bmatrix}
\text{var}(\bar{x}_1) & \text{cov}(\bar{x}_1, \bar{x}_2) & \text{cov}(\bar{x}_1, \bar{x}_3) & & & \\
\text{cov}(\bar{x}_1, \bar{x}_2) & \text{var}(\bar{x}_2) & \text{cov}(\bar{x}_2, \bar{x}_3) & & 0 & \\
\text{cov}(\bar{x}_1, \bar{x}_3) & \text{cov}(\bar{x}_2, \bar{x}_3) & \text{var}(\bar{x}_3) & & & \\
& & & \text{var}(\bar{y}_1) & \text{cov}(\bar{y}_1, \bar{y}_2) & \text{cov}(\bar{y}_1, \bar{y}_3) \\
& 0 & & \text{cov}(\bar{y}_1, \bar{y}_2) & \text{var}(\bar{y}_2) & \text{cov}(\bar{y}_2, \bar{y}_3) \\
& & & \text{cov}(\bar{y}_1, \bar{y}_3) & \text{cov}(\bar{y}_2, \bar{y}_3) & \text{var}(\bar{y}_3)
\end{bmatrix}
\tag{6.39}
$$

where

$$
\text{var}(\bar{x}_i) = \frac{\sum\limits_{j=1}^{l_i} (\bar{x}_{ij} - \bar{x}_i)^2}{l_i(l_i - 1)} = \frac{s_{x_i}^2}{l_i}
$$

and

$$
\text{cov}(\bar{x}_i, \bar{x}_{i'}) = \frac{\sum\limits_{j=1}^{l_i} (x_{ij} - \bar{x}_i)(x_{i'j} - \bar{x}_{i'})}{l_i(l_i - 1)}
$$

The orthogonal contrast corresponding to null hypothesis (6.36) is represented by the vector of coefficients (Table 6.7) as

$$
\underline{b} = \begin{bmatrix} 0 \\ 1 \\ -1 \\ 0 \\ -1 \\ 1 \end{bmatrix}
\tag{6.40}
$$

The d-test of null hypotheses (6.36) is then based on the vectors (6.38) and (6.40) and variance–covariance matrix (6.39) and of general form

$$
d = \frac{\underline{b}' \, \hat{\underline{\mu}}}{\sqrt{\underline{b}' \hat{\Sigma} \underline{b}}}
\tag{6.41}
$$

or specifically

$$
\begin{aligned}
d &= \frac{\bar{x}_2 - \bar{x}_3 - \bar{y}_2 + \bar{y}_3}{\sqrt{\text{var}(\bar{x}_2) + \text{var}(\bar{x}_3) + \text{var}(\bar{y}_2) + \text{var}(\bar{y}_3) - 2\,\text{cov}(\bar{x}_2, \bar{x}_3) - 2\,\text{cov}(\bar{y}_2, \bar{y}_3)}} \\
&= \frac{\bar{x}_2 - \bar{x}_3 - \bar{y}_2 + \bar{y}_3}{\sqrt{\dfrac{s_{x_2}^2 + s_{x_3}^2 - 2\,\text{cov}(x_2, x_3)}{l_c} + \dfrac{s_{y_2}^2 + s_{y_3}^2 - 2\,\text{cov}(y_2, y_3)}{l_T}}}
\end{aligned}
\tag{6.42}
$$

The second d-test in the sequence and a test of the null hypothesis (6.37) is again based on test statistic (6.41) but using the orthogonal contrast

$$\underset{\sim}{b} = \begin{bmatrix} 2 \\ -1 \\ -1 \\ -2 \\ 1 \\ 1 \end{bmatrix}$$

The d-test for null hypothesis (6.37) is

$$d = (2\bar{x}_1 - \bar{x}_2 - \bar{x}_3 - 2\bar{y}_1 + \bar{y}_2 + \bar{y}_3)$$
$$\times \left[\frac{4s_{x_1}^2 + s_{x_2}^2 + s_{x_3}^2 - 4\,\text{cov}(x_1, x_2) - 4\,\text{cov}(x_1, x_3) + 2\,\text{cov}(x_2\,x_3)}{l_c} \right.$$
$$\left. + \frac{4s_{y_1}^2 + s_{y_2}^2 + s_{y_3}^2 - 4\,\text{cov}(y_1, y_2) - 4\,\text{cov}(y_1, y_3) + 2\,\text{cov}(y_2, y_3)}{l_T} \right]^{-1/2}$$

Test statistic (6.41) generalizes for any number of sampling periods and contrasts (i.e., $\underset{\sim}{b}$) and is distributed as a t-statistic with $l_C + l_T - 2$ degrees of freedom.

By defining the elements of an orthogonal contrast ($\underset{\sim}{b}$) such that

$$\underset{\sim}{b} = \begin{bmatrix} b_{C1} \\ b_{C2} \\ \cdot \\ \cdot \\ b_{Ct} \\ b_{T1} \\ \cdot \\ \cdot \\ b_{Tt} \end{bmatrix}$$

the d-statistic can be written in the general form

$$d = \frac{\displaystyle\sum_{i=1}^{t} b_{Ci}\bar{x}_i + \sum_{i=1}^{t} b_{Ti}\bar{y}_i}{\sqrt{\displaystyle\sum_{i=1}^{t} b_{Ci}^2 \,\text{var}(\bar{x}_1) + \sum_{i=1}^{t} b_{Ti}^2 \,\text{var}(\bar{y}_i) + 2\sum_{i=1}^{t}\sum_{\substack{j=1 \\ i<j}}^{i} b_{Ci}b_{Cj} \,\text{cov}(\bar{x}_i, \bar{x}_j) + 2\sum_{i=1}^{t}\sum_{\substack{j=1 \\ i<j}}^{t} b_{Ti}b_{Tj} \,\text{cov}(\bar{y}_i, \bar{y}_j)}}$$
$$(6.43)$$

Test statistic (6.43) can be further simplified by implicitly incorporating the covariance terms rather than explicitly calculating each covariance in matrix (6.39). Define the contrasts

$$\hat{L}_{C_j} = \sum_{i=1}^{t} b_{C_i} x_{ij}$$

and

$$\hat{L}_{T_j} = \sum_{i=1}^{t} b_{T_i} y_{ij}$$

with the subsequent summary statistics

$$\hat{\bar{L}}_C = \sum_{j=1}^{l_C} \hat{L}_{C_j}/l_C$$

$$\hat{\bar{L}}_T = \sum_{j=1}^{l_T} \hat{L}_{T_j}/l_T$$

$$s_{\hat{L}_C}^2 = \frac{\sum_{j=1}^{l_C} (\hat{L}_{C_j} - \hat{\bar{L}}_C)^2}{(l_C - 1)}$$

$$s_{\hat{L}_T}^2 = \frac{\sum_{j=1}^{l_t} (\hat{L}_{T_j} - \hat{\bar{L}}_T)^2}{(l_T - 1)} .$$

Using these intermediate calculations, test statistic (6.43) can be rewritten as

$$d = \frac{\hat{\bar{L}}_C - \hat{\bar{L}}_T}{\sqrt{\dfrac{s_{\hat{L}_C}^2}{l_C} + \dfrac{s_{\hat{L}_T}^2}{l_T}}} . \qquad (6.44)$$

Choice of expressions (6.41), (6.43) or (6.44) for the d-test simply depends on the method of calculation. Typically, expression (6.44) will be easiest to compute. The value of statistics (6.41) and (6.43) is in the explicit expression of the variance structure that is useful in power calculations.

The contrasts associated with the sequence of tests of impact illustrated above are not the only set of contrasts that may be used in accident assessment. A forward sequence of tests (Table 6.7) also may be used to test for parallelism starting with sampling periods 1 and 2. Should significant effects be detected between periods 1 and 2, comparisons including subsequent sampling periods may be conducted until no significant difference in parallelism is found. Williams (1971), using monotone regression techniques, developed sequential t-tests to locate a threshold where response values level off. The technique of Williams (1971) compares the first versus the last treatment, and if significant, tests first versus next-to-last. Tables of critical values are presented whereby tests performed in the prescribed sequence have nominal size α, and actual size at most α making the test procedure conservative.

Example Analysis

Aerial spray programs to control insect pests have traditionally been used on apple orchards. Miscalculating the amount of pesticide added to the solvent can result in an accidental application of a pesticide at concentrations toxic to wildlife. For illustration, it will be assumed several orchards accidentally received a 100-fold increase in pesticide concentration. Five 1-ha treatment sites were established in oversprayed areas. An additional five 1-ha plots were established in orchards receiving prescribed pesticide levels after the application problem was identified.

The first of three population surveys were conducted 8 days after the accidental application of the wrong pesticide concentration. The remaining two survey periods were evenly spaced a month apart following the initial survey. Capture data from Lincoln Indices for the 30 population surveys are summarized in Table 6.8. The accident assessment will be based on tests of hypotheses (6.36) and (6.37).

The accident assessment provides 30 population surveys (Table 6.8) on which a test of effects can be based. Chi-square tests of homogeneity indicate capture probabilities were homogeneous for all populations within a survey period [period 1; $P(\chi^2_{18} > 16.8142) = 0.5359$; period 2, $P(\chi^2_{18} > 15.3496) = 0.6378$; period 3, $P(\chi^2_{18} > 10.673) = 0.9091$] but differed between periods [$P(\chi^2_{40} > 68.9703) = 0.0030$], constituting a case 2a type of homogeneity. From the results of the chi-square tests of homogeneity, assessment can be based on the catch indices (r) and linear contrasts calculated as a function of $\ln r$.

Test of parallelism for control and treatment sites between periods 2 and 3 [i.e., hypothesis (6.36)] is not rejected [$P(t_8 > |0.4082|) = 0.6938$] and supported by graphic evidence of mean abundance levels over time (Figure 6.11). Subsequent test of parallelism [i.e., hypothesis (6.37)], however, is rejected [$P(t_8 > |2.4395|) = 0.0406$], indicating a transitory impact between periods 1 and 2 (see Figure 6.11). Note that a simple test of the null hypothesis of equal mean abundance at control and treatment sites immediately after the accident would not have been rejected.

Designing Accident Assessment Studies

The power of test of parallelism in accident assessment will depend on the following: (1) test statistic employed, (2) α-level, (3) number of control (l_C) and treatment (l_T) sites, (4) capture probabilities (p_1, p_2), (5) magnitude of the impact and its transitory nature, (6) magnitude of spatial variance in abundance, and (7) temporal covariance in repeated measures of abundance. As with the odd's ratio design, power of the test depends on the number of sites used in the investigation and on the spatial variance among these sites.

Table 6.8

Capture Data, Abundance Estimates (3.5), and Estimates of Linear Contrasts ($\hat{L}_1 = x_2 - x_3$; $\hat{L}_2 = 2x_1 - x_2 - x_3$) for *Peromyscus leucopus* at Five Control and Five Treatment Sites in an Assessment of Effects of an Accidental Pesticide Overdose on Local Mammal Abundance*

Treatment designation	Replicate	Period 1					Period 2					Period 3					$\hat{L}_1$	$\hat{L}_2$
		n_1	n_2	m	r	$\hat{N}$	n_1	n_2	m	r	$\hat{N}$	n_1	n_2	m	r	$\hat{N}$		
Controls	1	33	22	12	43	59.15	60	59	38	81	92.85	31	31	13	49	72.14	0.5026	−0.7639
	2	23	33	12	44	61.77	62	71	42	91	104.49	33	39	15	57	84.00	0.4678	−0.9855
	3	24	28	13	39	50.79	49	48	30	67	78.03	30	35	16	59	64.65	0.1272	−0.9551
	4	27	17	9	35	49.40	42	58	29	71	83.57	30	29	13	46	65.43	0.4340	−0.9806
	5	25	20	10	35	48.64	53	48	32	69	79.18	22	25	8	39	65.44	0.5705	−0.7870
Treatment	1	25	25	11	39	55.33	64	64	37	91	110.18	31	29	9	51	91.80	0.5790	−1.1156
	2	31	25	15	41	51.00	63	63	36	90	109.70	42	32	16	58	80.53	0.4394	−1.1331
	3	23	24	9	38	59.00	56	49	28	77	97.28	24	28	8	44	79.56	0.5596	−0.8528
	4	20	23	10	33	44.82	48	56	31	73	86.28	40	44	19	65	91.25	0.1161	−1.4718
	5	29	31	15	45	59.00	65	61	34	92	115.91	49	52	22	79	114.22	0.1523	−1.2779
Tests of homogeneity					$\chi^2_{18} = 16.8142$					$\chi^2_{18} = 15.3496$						$\chi^2_{18} = 10.6373$		

*Results of tests of homogeneity on a periodwise basis are reported.

206

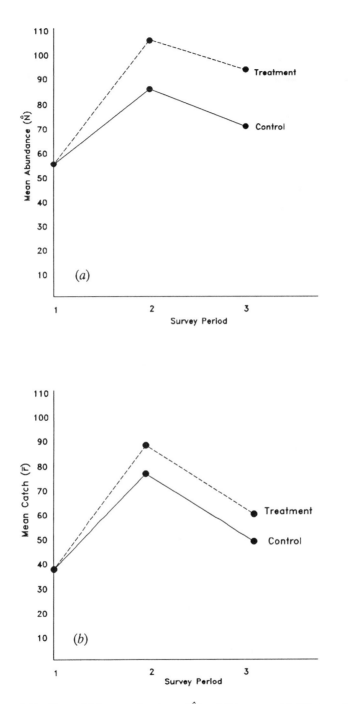

Figure 6.11 Plots of (a) mean abundance $(\hat{N})$ and (b) mean catch $(\bar{r})$ at control and treatment sites during three survey periods in an accident assessment.

The power of the tests of parallelism (6.43) can be computed from tables of noncentral F-distribution (Appendixces 2 and 3) using noncentrality parameter

$$\phi = \frac{1}{\sqrt{2}} \cdot \frac{\sqrt{l}\,|\underline{b}'\underline{\mu}|}{\sqrt{\sigma_{EX}^2}} \tag{6.45}$$

with 1 and $2(l-1)$ degrees of freedom for a balanced design. The vector $\underline{\mu}$ is the expected value of the vector of sample means (6.38) and σ_{EX}^2, the error variance under the alternative hypothesis of an impact. Vector $\underline{\mu}$ in all cases is expressed as

$$\underline{\mu} = \begin{bmatrix} \ln \mu_{C1} \\ \ln \mu_{C2} \\ \vdots \\ \ln \mu_{Ct} \\ \ln \mu_{T1} \\ \vdots \\ \ln \mu_{Tt} \end{bmatrix}$$

Bias corrections to $\underline{\mu}$ may be made if the form of the bias is known as in the case of the Lincoln Index.

Power Calculations of Case 1   Then capture probabilities are heterogeneous among abundance surveys compared, the error variance must reflect the mark–recapture sampling error induced by the surveys. The error variance under ln-transformed Lincoln Indices is approximately

$$\sigma_{EX}^2 = \sum_{t=1}^{t} b_{Ci}^2 \left[\frac{q_1 q_2}{p_1 p_2} \left(\frac{1}{\mu_{Ci}} + \frac{\sigma_{Ci}^2}{\mu_{Ci}^3} \right) + \sigma_{Ci}^2 \left(\frac{1}{\mu_{Ci}} + \frac{\sigma_{Ci}^2}{\mu_{Ci}^3} \right)^2 \right]$$
$$+ \sum_{i=1}^{t} b_{Ti}^2 \left[\frac{q_1 q_2}{p_1 p_2} \left(\frac{1}{\mu_{Ti}} + \frac{\sigma_{Ti}^2}{\mu_{Ti}^3} \right) + \sigma_{Ti}^2 \left(\frac{1}{\mu_{Ti}} + \frac{\sigma_{Ti}^2}{\mu_{Ti}^3} \right)^2 \right]$$
$$+ 2 \sum_{i=1}^{t} \sum_{\substack{j=1 \\ i<j}}^{t} b_{Ci} b_{Cj} \text{Cov}(N_{Ci}, N_{Cj}) \left(\frac{1}{\mu_{Ci}} + \frac{\sigma_{Ci}^2}{\mu_{Ci}^3} \right) \left(\frac{1}{\mu_{Cj}} + \frac{\sigma_{Cj}^2}{\mu_{Cj}^3} \right)$$
$$+ 2 \sum_{i=1}^{t} \sum_{\substack{j=1 \\ i<j}}^{t} b_{Ti} b_{Tj} \text{Cov}(N_{Ti}, N_{Ti}) \left(\frac{1}{\mu_{Ti}} + \frac{\sigma_{Ti}^2}{\mu_{Ti}^3} \right) \left(\frac{1}{\mu_{Tj}} + \frac{\sigma_{Tj}^2}{\mu_{Tj}^3} \right)$$

where p_1 and p_2 are the per-period capture probabilities that for simplicity are assumed common for all surveys.

Power Calculations for Case 2b   When the capture probablities are homogeneous among all population surveys compared, catch indices (r values) from the Lincoln Indices can be used to test for impact, and in that case, the error variance can be expressed as

$$
\sigma_{EX}^2 = \sum_{i=1}^{t} b_{Ci}^2 \left[\frac{q_1 q_2}{1 - q_1 q_2} \left(\frac{1}{\mu_{Ci}} + \frac{\sigma_{Ci}^2}{\mu_{Ci}^3} \right) + \sigma_{Ci}^2 \left(\frac{1}{\mu_{Ci}} + \frac{\sigma_{Ci}^2}{\mu_{Ci}^3} \right)^2 \right]
$$

$$
+ \sum_{i=1}^{t} b_{Ti}^2 \left[\frac{q_1 q_2}{1 - q_1 q_2} \left(\frac{1}{\mu_{Ti}} + \frac{\sigma_{Ti}^2}{\mu_{Ti}^3} \right) + \sigma_{Ti}^2 \left(\frac{1}{\mu_{Ti}} + \frac{\sigma_{Ti}^2}{\mu_{Ti}^3} \right)^2 \right]
$$

$$
+ 2 \sum_{i=1j=1}^{t} \sum_{i<j}^{t} b_{Ci} b_{Cj} \, \text{Cov}(N_{Ci}, N_{Cj}) \left(\frac{1}{\mu_{Ci}} + \frac{\sigma_{Ci}^2}{\mu_{Ci}^3} \right) \left(\frac{1}{\mu_{Cj}} + \frac{\sigma_{Cj}^2}{\mu_{Cj}^3} \right)
$$

$$
+ 2 \sum_{i=1j=1}^{t} \sum_{i<j}^{t} b_{Ti} b_{Tj} \, \text{Cov}(N_{Ti}, N_{Tj}) \left(\frac{1}{\mu_{Ti}} + \frac{\sigma_{Ti}^2}{\mu_{Ti}^3} \right) \left(\frac{1}{\mu_{Tj}} + \frac{\sigma_{Tj}^2}{\mu_{Tj}^3} \right).
$$

Generalization of Power Calculations   For survey techniques other than the Lincoln Index, expressions for the error variance in a test of parallelism from accident assessment can be written as functions of the average sampling error. In case 1 analysis, the error variance can be written as

$$
\sigma_{EX}^2 = \sum_{i=1}^{t} b_{Ci}^2 \left[\overline{(\text{Var}(\hat{N}_{Ci})} + \sigma_{Ci}^2 \left(\frac{1}{\mu_{Ci}} + \frac{\sigma_{Ci}^2}{\mu_{Ci}^3} \right)^2 \right]
$$

$$
+ \sum_{i=1}^{t} b_{Tj}^2 \left[\overline{(\text{Var}(\hat{N}_{Tj})} + \sigma_{Ti}^2 \left(\frac{1}{\mu_{Ti}} + \frac{\sigma_{Ti}^2}{\mu_{Ti}^3} \right)^2 \right]
$$

$$
+ 2 \sum_{i=1j=1}^{t} \sum_{i<j}^{t} b_{Ci} b_{Cj} \, \text{Cov}(N_{Ci}, N_{Cj}) \left(\frac{1}{\mu_{Ci}} + \frac{\sigma_{Ci}^2}{\mu_{Ci}^3} \right) \left(\frac{1}{\mu_{Cj}} + \frac{\sigma_{Cj}^2}{\mu_{Cj}^3} \right)
$$

$$
+ 2 \sum_{i=1j=1}^{t} \sum_{i<j}^{t} b_{Ti} b_{Tj} \, \text{Cov}(N_{Ti}, N_{Tj}) \left(\frac{1}{\mu_{Ti}} + \frac{\sigma_{Ti}^2}{\mu_{Ti}^3} \right) \left(\frac{1}{\mu_{Tj}} + \frac{\sigma_{Tj}^2}{\mu_{Tj}^3} \right) \tag{6.46}
$$

for use with noncentrality parameter (6.45).

When catch indices can be used to test for impact under case 2b homogeneity, the error variance for use with any survey model and noncentrality parameter (6.45) with a balanced design is expressed as

$$
\sigma_{EX}^2 = \sum_{i=1}^{t} b_{Ci}^2 \left[\frac{1 - P}{P} \left(\frac{1}{\mu_{Ci}} + \frac{\sigma_{Ci}^2}{\mu_{Ci}^3} \right) + \sigma_{Ci}^2 \left(\frac{1}{\mu_{Ci}} + \frac{\sigma_{Ci}^2}{\mu_{Ci}^3} \right)^2 \right]
$$

$$
+ \sum_{i=1}^{t} b_{Ti}^2 \left[\frac{1 - P}{P} \left(\frac{1}{\mu_{Ti}} + \frac{\sigma_{Ti}^2}{\mu_{Ti}^3} \right) + \sigma_{Ti}^2 \left(\frac{1}{\mu_{Ti}} + \frac{\sigma_{Ti}^2}{\mu_{Ti}^3} \right)^2 \right]
$$

$$
+ 2 \sum_{i=1 j=1}^{t} \sum_{i<j}^{t} b_{Ci} b_{Cj} \, \text{Cov}(N_{Ci}, N_{Cj}) \left(\frac{1}{\mu_{Ci}} + \frac{\sigma_{Ci}^2}{\mu_{Ci}^3} \right) \left(\frac{1}{\mu_{Cj}} + \frac{\sigma_{Cj}^2}{\mu_{Cj}^3} \right)
$$

$$
+ 2 \sum_{i=1 j=1}^{t} \sum_{i<j}^{t} b_{Ti} b_{Tj} \, \text{Cov}(N_{Ti}, N_{Tj}) \left(\frac{1}{\mu_{Ti}} + \frac{\sigma_{Ti}^2}{\mu_{Ti}^3} \right) \left(\frac{1}{\mu_{Tj}} + \frac{\sigma_{Tj}^2}{\mu_{Tj}^3} \right)
$$

where P is the overall probability of capture during the mark–recapture surveys.

EXAMPLE In the aerial spray example, the test of the null hypothesis of parallelism between periods 2 and 3 was not rejected. Because the sequential testing has implications with regard to making inferences to the duration of the impact, *post hoc* power calculation will be performed to interpret the nonrejection of hypothesis (6.36). Table 6.9 summarizes parameter estimates necessary in computing the power of the test.

Using the parameter estimates from Table 6.9, the noncentrality parameter associated with the test of parallelism between periods 2 and 3 in the aerial spray study is computed to be

$$\phi_{1.8} = \frac{1}{\sqrt{2}} \cdot \frac{\sqrt{5}\,|0.4202 - 0.3693|}{\sqrt{0.01833}} = 0.5972,$$

where

$$
\begin{aligned}
\sigma^2_{\text{EX}} = {} & (1)^2\left[\frac{(1-0.83)}{0.83}\left(\frac{1}{87.62} + \frac{93.72}{(87.62)^3}\right) + 93.72\left(\frac{1}{87.62} + \frac{93.72}{(87.62)^3}\right)^2\right] \\
& + (-1)^2\left[\frac{(1-0.66)}{0.66}\left(\frac{1}{70.33} + \frac{70.33}{(70.33)^3}\right) + 70.33\left(\frac{1}{70.33} + \frac{70.33}{(70.33)^3}\right)^2\right] \\
& + (-1)^2\left[\frac{(1-0.83)}{0.83}\left(\frac{1}{103.87} + \frac{84.69}{(103.87)^3}\right) + 84.69\left(\frac{1}{103.87} + \frac{84.69}{(103.87)^3}\right)^2\right] \\
& + (1)^2\left[\frac{(1-0.66)}{0.66}\left(\frac{1}{91.47} + \frac{91.47}{(91.47)^3}\right) + 91.47\left(\frac{1}{91.47} + \frac{91.47}{(91.47)^3}\right)^2\right] \\
& + 2(1(-1))88.91\left(\frac{1}{87.62} + \frac{93.72}{(87.62)^3}\right)\left(\frac{1}{70.33} + \frac{70.33}{(70.33)^3}\right) \\
& + 2(1(-1))73.67\left(\frac{1}{103.87} + \frac{84.69}{(103.87)^3}\right)\left(\frac{1}{91.47} + \frac{91.47}{(91.47)^3}\right) \\
= {} & 0.01833.
\end{aligned}
$$

For a two-tailed test at $\alpha = 0.10$, the aerial spray program had a power of 0.20 to reject the null hypothesis of parallelism between periods 2 and 3. Increasing the level of field

Table 6.9

Estimated Population Parameters Used in post hoc Power Calculations
on Accident Assessment

Mean abundance (μ_N)	Overall capture probability	Population	Variance–covariance matrix			
			N_{C2}	N_{C3}	N_{T2}	N_{T3}
87.62	0.83	N_{C2}	93.72	88.91	0	0
70.33	0.66	N_{C3}		70.33*	0	0
103.87	0.83	N_{T2}			84.69	73.67
91.47	0.66	N_{T3}				91.47*

* Variance component estimation found $\hat{\sigma}^2_N < 0$; hence, set $\hat{\sigma}^2_N = \hat{\mu}_N$.

replication to $l = 20$ sites per treatment from the original effort of $l = 5$ does little to improve the power of the test, i.e., $1 - \beta = 0.50(\phi_{1,38} = 1.1840)$. Inspection of Figure 6.10 suggests good agreement with the null hypothesis of parallelism and is an explanation of the low power to detect effects of the spray beyond period 2.

Design Recommendations

Impact and impact assessment studies are among the most difficult of population investigations to properly design and analyze. Constraints on use of experimental principles of randomization and replication require that inferences be model-dependent. To enhance validity of the conclusion, impact studies typically must include a temporal dimension to their design. Consequently, not only are design principles less understood, but environmental assessments often must include design elements that can be avoided in manipulative experiments.

A key component in impact and impact assessment designs presented has been the use of proportional abundance estimators $(\hat{K})$. In odd's ratio design, the linear contrast (6.2) can be written as

$$\ln D_i = \ln \left(\frac{N_{C0_i}}{N_{T0_i}} \right) - \ln \left(\frac{N_{C1_i}}{N_{T1_i}} \right)$$

$$\ln D_i = \ln K_{0_i} - \ln K_{1_i}. \tag{6.47}$$

In this form, the value of $\ln D_i$ can be seen to be the ln-difference in proportional abundance between pre- and posttreatment periods of the study. Expression (6.47) is analogous to the numerator of the F-test (6.21) for the impact assessment hypothesis (6.20). The linear contrast (6.34) for a two-period accident assessment is also of the general form (6.47). In both the odd's ratio and CTP designs, estimates of K are used to eliminate confounding effects assumed to have a multiplicative influence on population abundance.

For odd's ratio design, elimination of effects of plot location and survey period by the calculation of proportional abundance provides the basis for statistical inference to a treatment effect. However, the longer the duration between pre-and posttreatment surveys, the less likely it is that temporal effects are eliminated by $\hat{K}$, and inferences to treatment effects become weaker. Elimination of plot effects is also accomplished through model-dependent assumptions that are only approximately true.

In CTP design, use of proportional abundance has a dual purpose. The first purpose is the same as in the odd's ratio design, that of eliminating temporal effects that may be confounded with effects of impact. The second purpose of using $\hat{K}$ is to provide repeated observations of a factorial treatment condition so that $\hat{\sigma}^2_{EX}$ can be estimated for testing hypotheses. Annual observations of $\hat{K}$ from a site pair are viewed as replicates of a factorial treatment condition defined by location × season × monitoring phase. It is for this reason that the power calculations of a CTP design use temporal moments in animal abundance rather than spatial variance used in manipulative experiments.

Repeated observations of abundance from a population are not typically independent. To mimic this correlation, simulation models were developed (Skalski 1985a) to induce an autocorrelation among abundance estimates. Despite such correlation, test statistic (6.21) remained F-distributed. The reason is the elimination of autocorrelation by the use of proportional abundance estimators in data analysis. Actual field data show a similar elimination of autocorrelation when proportional abundance is estimated (McKenzie et al. 1977). This absence of autocorrelation is a benefit of CTP designs not encountered in other impact assessment designs (Green 1979).

In analysis of environmental accidents, ln-linear contrasts are used to eliminate potentially confounding influences of site and survey period from the effects of impact. Because site pairing was not employed, autocorrelation existed among the repeated estimates of abundance, requiring test statistic (6.43) to explicitly include covariances in abundance in the analysis. Central limit theorem ensures approximate normality of the contrasts as sample sizes increase.

Recommendations for design of impact and impact assessment studies include the pertinent remarks made at the end of Chapter 5. Additional considerations unique to designing constrained environmental investigations include

1. Identification of constraints imposed by the investigation with regard to randomization and replication (Figure 1.2)
2. Incorporation of all prior knowledge as to where, when, and how the impact is to occur into the design of the field investigation
3. Specification of the statistical tests of impact using response models to determine model-dependent assumptions
4. Expression of the impact hypothesis in statistical terms as a function of model parameters
5. Use of a preliminary survey that is consistent with the objective of the consummate field design in order to estimate variance components for sample size calculations
6. Evaluation of economic and inferential costs of conducting a constrained investigation relative to other design options
7. Establishment of a field design whose spatial and temporal dimensions permit model-dependent estimates of effects of impact

8. Determination of the validity of model-dependent assumptions (e.g., tests of additivity)
9. Where possible, conducting auxiliary investigations of waste process flow or stresses to provide ancillary data for establishing cause–effect relationship

With these steps incorporated in monitoring studies, environmental assessment may retain many of the inferential capabilities of manipulative experiments.

Results of design optimization also can provide objective criteria for evaluating monitoring activities. Identification of an effective and sound environmental assessment design is not always sufficient. High costs of an environmental assessment almost necessitate that a field study be efficient before it can be effective. When an optimal design is found to possess insufficient power to detect a biologically important change for a given budget, all alternative allocations of effort for the same budget are likewise shown to be ineffectual. As such, results of optimization can either substantiate the value of a species for impact assessment or provide sound statistical, biological, and economic reasons why such monitoring effort is not feasible (Skalski 1984).

Organisms within the waste process flow that provide the greatest power to detect impacts for fixed costs are preferred. Optimization procedures as outlined above can be used to identify the more promising species based on costs of sampling and variation in population response to a stressor. Biomonitors with little or no chance of conveying an impact can be identified through the process of design optimization and can be eliminated in favor of more sensitive, less variable, and more easily sampled organisms.

Alternatively, for commercially, politically, or ecologically important species that must be monitored for impact, the optimization procedures can be used to determine the minimal budget needed to ensure significant impacts are detected. Through identification of the most cost-effective design (i.e., design with minimum C_0 for fixed power), research budgets can be adjusted to provide sufficient resources for the task or identify a potentially unsuccessful effort. In either case, monitoring investigations can be made accountable for the design decisions that may ultimately influence resource management decisions.

1

General Variance Component Formula

If X and Y are jointly distributed random variables, then the identity

$$Y = E(Y) + E(Y|X) - E(Y) + Y - E(Y|X) \qquad \text{(A.1)}$$

leads to the decomposition of the variance of Y into two components

$$\text{Var}(Y) = \text{Var}[E(Y|X)] + E[\text{Var}(Y|X)]$$

representing, respectively, the variance of the conditional mean of Y given X and the average conditional variance of Y given X (assuming the existence of these conditional means and variances).

This well-known fundamental result is obtained by subtracting the unconditional mean $E(Y)$ from both sides of (A.1), then squaring both sides and calculating the mathematical expectation of both sides in two stages. First, the conditional expectation is taken with respect to the conditional distribution of Y given X, and then expectation is taken with respect to the marginal distribution of X. Operating in this manner on the left-hand side, one immediately obtains $\text{Var}(Y)$, using the fact that the expectation of a conditional expectation (assuming the latter exists) is the unconditional expectation. Thus, just as $E\{E(Y|X)\} = E(Y)$, one obtains

$$E(E\{[Y - E(Y)]^2|X\}) = E[Y - E(Y)]^2 = \text{Var}(Y).$$

Squaring and expanding the right-hand side and then operating in this two-stage manner, one obtains

$$E(E\{[(Y\,|\,X) - E(Y)]^2\,|\,X\}) + E(E\{[Y - E(Y\,|\,X)]^2\,|\,X\})$$
$$+ 2E(E\{[E(Y\,|\,X) - E(Y)][Y - E(Y\,|\,X)]\,|\,X\}).$$

Noting that $[E(Y\,|\,X) - E(Y)]$ depends only on X and not on Y, we obtain the somewhat simpler form

$$\mathrm{Var}(Y) = E([E(Y\,|\,X) - E(Y)]^2) + E(E\{[Y - E(Y\,|\,X)]^2\,|\,X\})$$
$$+ 2E([E(Y\,|\,X) - E(Y)]E\{[Y - E(Y\,|\,X)]\,|\,X\}).$$

Finally, noting that

$$E\{[Y - E(Y\,|\,X)]\,|\,X\} = 0$$
$$E\{[Y - E(Y\,|\,X)]^2\,|\,X\} = \mathrm{Var}(Y\,|\,X)$$
$$E([E(Y\,|\,X) - E(Y)]^2) = \mathrm{Var}[E(Y\,|\,X)]$$

one obtains the general variance component formula

$$\mathrm{Var}(Y) = \mathrm{Var}[E(Y\,|\,X)] + E[\mathrm{Var}(Y\,|\,X)].$$

Noncentral F-Tables

Selected tables for the noncentral F-distribution with 1 and f_2 degrees of freedom. Values of $1 - \beta$ (i.e., Power) are given as a function of Φ and f_2 for values of $\alpha = 0.02, 0.05, 0.10,$ and 0.20. Values of $1 - \beta$ are associated with two-tailed tests at α or one-tailed tests at $\alpha/2$.

α = 0.02

f_2 \ $\phi=$	3.0	2.6	2.2	2.0	1.8	1.6	1.4	1.3	1.2	1.1	1.0	0.9	0.8	0.7	0.6	0.5	0.4	0.3	0.2	0.1
1	0.0202	0.0208	0.0218	0.0231	0.0248	0.0268	0.0291	0.0316	0.0343	0.0372	0.0403	0.0434	0.0467	0.0500	0.0569	0.0638	0.0708	0.0779	0.0919	0.1060
2	0.0204	0.0216	0.0235	0.0262	0.0297	0.0339	0.0388	0.0445	0.0509	0.0580	0.0659	0.0743	0.0834	0.0932	0.1145	0.1380	0.1636	0.1909	0.2502	0.3138
3	0.0205	0.0222	0.0249	0.0287	0.0337	0.0399	0.0473	0.0560	0.0660	0.0774	0.0901	0.1042	0.1197	0.1366	0.1743	0.2170	0.2641	0.3149	0.4232	0.5337
4	0.0207	0.0226	0.0259	0.0306	0.0368	0.0446	0.0540	0.0651	0.0781	0.0929	0.1098	0.1286	0.1494	0.1722	0.2236	0.2818	0.3455	0.4130	0.5512	0.6805
5	0.0207	0.0230	0.0267	0.0321	0.0392	0.0481	0.0591	0.0721	0.0874	0.1050	0.1250	0.1475	0.1725	0.1999	0.2615	0.3309	0.4060	0.4839	0.6366	0.7681
6	0.0208	0.0232	0.0273	0.0332	0.0410	0.0509	0.0630	0.0775	0.0946	0.1144	0.1369	0.1623	0.1905	0.2213	0.2906	0.3680	0.4507	0.5349	0.6938	0.8213
7	0.0209	0.0234	0.0278	0.0341	0.0425	0.0531	0.0661	0.0818	0.1004	0.1218	0.1464	0.1740	0.2046	0.2382	0.3133	0.3965	0.4843	0.5725	0.7334	0.8553
8	0.0209	0.0236	0.0282	0.0348	0.0436	0.0548	0.0687	0.0853	0.1050	0.1278	0.1540	0.1834	0.2160	0.2517	0.3312	0.4188	0.5103	0.6009	0.7619	0.8781
9	0.0209	0.0237	0.0285	0.0354	0.0446	0.0563	0.0707	0.0882	0.1088	0.1328	0.1602	0.1911	0.2253	0.2627	0.3457	0.4367	0.5308	0.6229	0.7831	0.8942
10	0.0210	0.0238	0.0288	0.0359	0.0454	0.0575	0.0725	0.0906	0.1120	0.1369	0.1654	0.1975	0.2330	0.2718	0.3577	0.4512	0.5473	0.6405	0.7994	0.9060
12	0.0210	0.0240	0.0292	0.0366	0.0466	0.0594	0.0752	0.0943	0.1170	0.1434	0.1735	0.2075	0.2450	0.2859	0.3761	0.4734	0.5722	0.6664	0.8225	0.9219
14	0.0210	0.0241	0.0295	0.0372	0.0475	0.0608	0.0772	0.0971	0.1207	0.1482	0.1796	0.2149	0.2540	0.2964	0.3896	0.4895	0.5899	0.6846	0.8380	0.9320
16	0.0211	0.0242	0.0297	0.0376	0.0483	0.0619	0.0788	0.0993	0.1236	0.1519	0.1843	0.2207	0.2608	0.3042	0.3999	0.5016	0.6032	0.6980	0.8490	0.9388
18	0.0211	0.0243	0.0299	0.0380	0.0488	0.0628	0.0801	0.1010	0.1259	0.1549	0.1880	0.2252	0.2663	0.3108	0.4080	0.5111	0.6134	0.7083	0.8573	0.9437
20	0.0211	0.0244	0.0301	0.0383	0.0493	0.0635	0.0811	0.1025	0.1278	0.1573	0.1911	0.2290	0.2707	0.3159	0.4145	0.5187	0.6216	0.7164	0.8636	0.9474
22	0.0211	0.0244	0.0302	0.0385	0.0497	0.0641	0.0819	0.1036	0.1294	0.1594	0.1936	0.2320	0.2744	0.3202	0.4199	0.5249	0.6282	0.7230	0.8687	0.9503
24	0.0211	0.0245	0.0303	0.0387	0.0500	0.0646	0.0827	0.1046	0.1307	0.1611	0.1957	0.2346	0.2774	0.3238	0.4244	0.5301	0.6337	0.7283	0.8728	0.9526
26	0.0211	0.0245	0.0304	0.0389	0.0503	0.0650	0.0833	0.1055	0.1318	0.1625	0.1975	0.2368	0.2801	0.3268	0.4282	0.5345	0.6384	0.7329	0.8762	0.9545
28	0.0211	0.0246	0.0305	0.0390	0.0505	0.0654	0.0838	0.1062	0.1328	0.1638	0.1991	0.2387	0.2823	0.3294	0.4314	0.5383	0.6423	0.7367	0.8790	0.9560
30	0.0211	0.0246	0.0305	0.0392	0.0508	0.0657	0.0843	0.1069	0.1337	0.1649	0.2005	0.2404	0.2843	0.3317	0.4343	0.5415	0.6458	0.7400	0.8814	0.9573
35	0.0212	0.0247	0.0307	0.0394	0.0512	0.0663	0.0852	0.1082	0.1354	0.1671	0.2032	0.2437	0.2883	0.3363	0.4400	0.5480	0.6526	0.7466	0.8862	0.9599
40	0.0212	0.0247	0.0308	0.0396	0.0515	0.0668	0.0859	0.1091	0.1367	0.1688	0.2053	0.2463	0.2913	0.3398	0.4443	0.5529	0.6576	0.7514	0.8897	0.9617
50	0.0212	0.0248	0.0309	0.0399	0.0520	0.0675	0.0870	0.1105	0.1385	0.1711	0.2083	0.2499	0.2955	0.3446	0.4503	0.5597	0.6647	0.7581	0.8944	0.9641
60	0.0212	0.0248	0.0310	0.0401	0.0523	0.0680	0.0876	0.1115	0.1398	0.1727	0.2103	0.2523	0.2984	0.3479	0.4544	0.5643	0.6693	0.7625	0.8974	0.9656
80	0.0212	0.0249	0.0312	0.0403	0.0527	0.0686	0.0885	0.1127	0.1414	0.1748	0.2128	0.2553	0.3019	0.3520	0.4594	0.5699	0.6751	0.7679	0.9011	0.9674
100	0.0212	0.0249	0.0312	0.0405	0.0529	0.0690	0.0890	0.1134	0.1423	0.1760	0.2143	0.2572	0.3041	0.3545	0.4625	0.5733	0.6786	0.7712	0.9033	0.9684
120	0.0212	0.0249	0.0313	0.0406	0.0531	0.0692	0.0894	0.1141	0.1430	0.1768	0.2153	0.2584	0.3055	0.3562	0.4645	0.5755	0.6809	0.7733	0.9047	0.9691
200	0.0212	0.0250	0.0314	0.0408	0.0534	0.0697	0.0901	0.1148	0.1442	0.1784	0.2174	0.2609	0.3084	0.3595	0.4686	0.5800	0.6854	0.7775	0.9075	0.9705

α = 0.05

f_2 \ φ =	0.1	0.2	0.3	0.4	0.5	0.6	0.7	0.8	0.9	1.0	1.1	1.2	1.3	1.4	1.6	1.8	2.0	2.2	2.6	3.0
1	0.0505	0.0520	0.0544	0.0578	0.0619	0.0669	0.0726	0.0788	0.0856	0.0928	0.1004	0.1082	0.1163	0.1246	0.1414	0.1585	0.1757	0.1929	0.2270	0.2608
2	0.0509	0.0537	0.0583	0.0647	0.0729	0.0828	0.0943	0.1075	0.1221	0.1383	0.1557	0.1744	0.1943	0.2153	0.2598	0.3073	0.3568	0.4074	0.5085	0.6050
3	0.0512	0.0549	0.0610	0.0696	0.0806	0.0941	0.1101	0.1284	0.1491	0.1721	0.1973	0.2245	0.2535	0.2843	0.3498	0.4192	0.4900	0.5602	0.6904	0.7978
4	0.0514	0.0557	0.0628	0.0729	0.0859	0.1018	0.1208	0.1427	0.1676	0.1952	0.2256	0.2585	0.2935	0.3306	0.4090	0.4905	0.5716	0.6491	0.7831	0.8802
5	0.0516	0.0563	0.0641	0.0752	0.0896	0.1073	0.1283	0.1528	0.1805	0.2114	0.2453	0.2820	0.3211	0.3622	0.4485	0.5368	0.6228	0.7026	0.8330	0.9186
6	0.0517	0.0567	0.0651	0.0769	0.0923	0.1113	0.1339	0.1601	0.1899	0.2232	0.2597	0.2990	0.3408	0.3847	0.4762	0.5685	0.6569	0.7371	0.8626	0.9389
7	0.0517	0.0570	0.0658	0.0782	0.0943	0.1143	0.1381	0.1657	0.1971	0.2321	0.2704	0.3117	0.3556	0.4014	0.4964	0.5913	0.6808	0.7607	0.8816	0.9510
8	0.0518	0.0572	0.0663	0.0792	0.0960	0.1166	0.1413	0.1700	0.2027	0.2390	0.2788	0.3216	0.3670	0.4142	0.5117	0.6083	0.6985	0.7777	0.8946	0.9587
9	0.0519	0.0574	0.0668	0.0800	0.0973	0.1186	0.1440	0.1735	0.2071	0.2445	0.2855	0.3294	0.3760	0.4243	0.5238	0.6215	0.7119	0.7904	0.9039	0.9640
10	0.0519	0.0576	0.0672	0.0807	0.0983	0.1201	0.1461	0.1764	0.2108	0.2491	0.2909	0.3358	0.3833	0.4325	0.5334	0.6320	0.7225	0.8003	0.9110	0.9678
12	0.0520	0.0578	0.0677	0.0817	0.1000	0.1225	0.1495	0.1808	0.2164	0.2560	0.2992	0.3455	0.3944	0.4449	0.5479	0.6476	0.7380	0.8146	0.9207	0.9728
14	0.0520	0.0580	0.0681	0.0825	0.1012	0.1243	0.1519	0.1840	0.2205	0.2610	0.3053	0.3526	0.4024	0.4538	0.5582	0.6586	0.7487	0.8244	0.9272	0.9760
16	0.0520	0.0582	0.0685	0.0831	0.1021	0.1256	0.1538	0.1865	0.2236	0.2649	0.3098	0.3579	0.4085	0.4606	0.5659	0.6667	0.7567	0.8315	0.9317	0.9781
18	0.0521	0.0583	0.0687	0.0835	0.1028	0.1267	0.1552	0.1884	0.2261	0.2679	0.3135	0.3621	0.4132	0.4658	0.5719	0.6730	0.7627	0.8369	0.9351	0.9797
20	0.0521	0.0584	0.0689	0.0839	0.1034	0.1276	0.1564	0.1900	0.2281	0.2703	0.3164	0.3655	0.4170	0.4700	0.5767	0.6780	0.7675	0.8411	0.9377	0.9808
22	0.0521	0.0584	0.0691	0.0842	0.1039	0.1283	0.1574	0.1913	0.2297	0.2723	0.3187	0.3683	0.4201	0.4735	0.5806	0.6820	0.7713	0.8445	0.9397	0.9817
24	0.0521	0.0585	0.0692	0.0845	0.1043	0.1289	0.1582	0.1924	0.2311	0.2740	0.3207	0.3706	0.4228	0.4763	0.5838	0.6854	0.7745	0.8472	0.9414	0.9825
26	0.0521	0.0586	0.0694	0.0847	0.1046	0.1294	0.1589	0.1933	0.2322	0.2754	0.3224	0.3726	0.4250	0.4788	0.5866	0.6882	0.7772	0.8496	0.9427	0.9831
28	0.0521	0.0586	0.0695	0.0849	0.1049	0.1298	0.1595	0.1941	0.2332	0.2767	0.3239	0.3742	0.4269	0.4809	0.5889	0.6906	0.7794	0.8515	0.9439	0.9835
30	0.0521	0.0586	0.0696	0.0850	0.1052	0.1302	0.1601	0.1948	0.2341	0.2777	0.3252	0.3757	0.4285	0.4827	0.5910	0.6927	0.7814	0.8532	0.9449	0.9840
35	0.0522	0.0587	0.0697	0.0854	0.1057	0.1310	0.1611	0.1961	0.2359	0.2799	0.3277	0.3786	0.4318	0.4863	0.5950	0.6968	0.7853	0.8565	0.9468	0.9848
40	0.0522	0.0588	0.0699	0.0856	0.1061	0.1315	0.1619	0.1972	0.2372	0.2815	0.3296	0.3809	0.4343	0.4890	0.5980	0.6999	0.7881	0.8590	0.9482	0.9853
50	0.0522	0.0589	0.0701	0.0860	0.1067	0.1324	0.1630	0.1987	0.2390	0.2838	0.3323	0.3840	0.4378	0.4928	0.6023	0.7042	0.7921	0.8624	0.9501	0.9861
60	0.0522	0.0589	0.0702	0.0862	0.1070	0.1329	0.1638	0.1997	0.2403	0.2853	0.3341	0.3860	0.4401	0.4953	0.6051	0.7070	0.7947	0.8646	0.9513	0.9866
80	0.0522	0.0590	0.0704	0.0865	0.1075	0.1336	0.1647	0.2009	0.2419	0.2872	0.3364	0.3886	0.4430	0.4985	0.6086	0.7105	0.7980	0.8673	0.9528	0.9872
100	0.0522	0.0590	0.0705	0.0867	0.1078	0.1340	0.1653	0.2016	0.2428	0.2884	0.3377	0.3902	0.4447	0.5004	0.6107	0.7126	0.7999	0.8689	0.9537	0.9875
120	0.0523	0.0591	0.0705	0.0868	0.1080	0.1343	0.1657	0.2021	0.2434	0.2891	0.3387	0.3912	0.4459	0.5017	0.6121	0.7140	0.8012	0.8700	0.9542	0.9877
200	0.0523	0.0591	0.0707	0.0870	0.1084	0.1348	0.1665	0.2031	0.2447	0.2907	0.3405	0.3933	0.4482	0.5042	0.6148	0.7168	0.8037	0.8721	0.9554	0.9882

$$\alpha = 0.10$$

f_2	$\phi =$ 0.1	0.2	0.3	0.4	0.5	0.6	0.7	0.8	0.9	1.0	1.1	1.2	1.3	1.4	1.6	1.8	2.0	2.2	2.6	3.0
1	0.1010	0.1039	0.1087	0.1153	0.1236	0.1333	0.1444	0.1567	0.1700	0.1840	0.1988	0.2140	0.2296	0.2455	0.2777	0.3100	0.3420	0.3736	0.4349	0.4931
2	0.1017	0.1068	0.1153	0.1269	0.1418	0.1595	0.1800	0.2030	0.2284	0.2557	0.2848	0.3154	0.3472	0.3798	0.4466	0.5137	0.5791	0.6412	0.7509	0.8372
3	0.1021	0.1085	0.1191	0.1338	0.1525	0.1749	0.2010	0.2303	0.2627	0.2977	0.3349	0.3739	0.4142	0.4554	0.5382	0.6187	0.6936	0.7606	0.8658	0.9330
4	0.1024	0.1096	0.1215	0.1380	0.1590	0.1843	0.2137	0.2469	0.2834	0.3227	0.3645	0.4081	0.4529	0.4983	0.5883	0.6734	0.7500	0.8154	0.9101	0.9625
5	0.1026	0.1103	0.1230	0.1408	0.1634	0.1906	0.2221	0.2577	0.2968	0.3389	0.3835	0.4299	0.4773	0.5250	0.6186	0.7055	0.7817	0.8448	0.9312	0.9745
6	0.1027	0.1108	0.1241	0.1427	0.1664	0.1949	0.2280	0.2653	0.3062	0.3502	0.3966	0.4445	0.4938	0.5430	0.6387	0.7262	0.8015	0.8627	0.9430	0.9806
7	0.1028	0.1111	0.1250	0.1442	0.1687	0.1982	0.2324	0.2708	0.3130	0.3584	0.4062	0.4556	0.5058	0.5559	0.6528	0.7405	0.8150	0.8745	0.9504	0.9841
8	0.1029	0.1114	0.1256	0.1453	0.1704	0.2006	0.2357	0.2751	0.3183	0.3647	0.4134	0.4637	0.9515	0.5656	0.6633	0.7510	0.8247	0.8828	0.9553	0.9863
9	0.1029	0.1116	0.1261	0.1462	0.1718	0.2026	0.2383	0.2784	0.3224	0.3696	0.4191	0.4701	0.5218	0.5731	0.6714	0.7590	0.8320	0.8889	0.9589	0.9879
10	0.1030	0.1118	0.1265	0.1469	0.1729	0.2042	0.2404	0.2812	0.3258	0.3735	0.4236	0.4752	0.5273	0.5791	0.6777	0.7652	0.8377	0.8937	0.9615	0.9890
12	0.1030	0.1121	0.1271	0.1480	0.1746	0.2066	0.2437	0.2853	0.3308	0.3795	0.4305	0.4829	0.5357	0.5880	0.6872	0.7744	0.8459	0.9005	0.9652	0.9904
14	0.1031	0.1123	0.1276	0.1488	0.1758	0.2084	0.2460	0.2882	0.3344	0.3838	0.4354	0.4884	0.5417	0.5943	0.6938	0.7808	0.8515	0.9051	0.9676	0.9914
16	0.1031	0.1124	0.1279	0.1494	0.1768	0.2097	0.2478	0.2905	0.3372	0.3870	0.4391	0.4925	0.5462	0.5990	0.6988	0.7855	0.8556	0.9084	0.9693	0.9920
18	0.1031	0.1126	0.1282	0.1499	0.1775	0.2107	0.2492	0.2922	0.3393	0.3896	0.4420	0.4957	0.5496	0.6027	0.7026	0.7891	0.8588	0.9109	0.9706	0.9925
20	0.1032	0.1127	0.1284	0.1503	0.1781	0.2116	0.2503	0.2937	0.3410	0.3916	0.4443	0.4983	0.5524	0.6056	0.7056	0.7919	0.8612	0.9129	0.9715	0.9928
22	0.1032	0.1127	0.1286	0.1506	0.1786	0.2123	0.2512	0.2948	0.3424	0.3932	0.4462	0.5004	0.5547	0.6080	0.7080	0.7943	0.8632	0.9144	0.9723	0.9931
24	0.1032	0.1128	0.1287	0.1508	0.1790	0.2128	0.2519	0.2958	0.3436	0.3946	0.4478	0.5021	0.5565	0.6100	0.7101	0.7962	0.8649	0.9157	0.9729	0.9933
26	0.1032	0.1129	0.1288	0.1511	0.1793	0.2133	0.2526	0.2966	0.3446	0.3958	0.4491	0.5036	0.5581	0.6116	0.7118	0.7978	0.8662	0.9168	0.9734	0.9935
28	0.1032	0.1129	0.1290	0.1513	0.1796	0.2137	0.2532	0.2973	0.3455	0.3968	0.4503	0.5049	0.5595	0.6131	0.7132	0.7991	0.8674	0.9177	0.9739	0.9937
30	0.1032	0.1129	0.1290	0.1514	0.1799	0.2141	0.2536	0.2979	0.3462	0.3977	0.4513	0.5060	0.5561	0.6143	0.7145	0.8003	0.8684	0.9185	0.9743	0.9938
35	0.1033	0.1130	0.1292	0.1518	0.1804	0.2148	0.2546	0.2991	0.3477	0.3994	0.4533	0.5082	0.5630	0.6168	0.7170	0.8026	0.8704	0.9200	0.9750	0.9940
40	0.1033	0.1131	0.1294	0.1520	0.1808	0.2154	0.2553	0.3001	0.3488	0.4007	0.4547	0.5098	0.5648	0.6186	0.7189	0.8044	0.8719	0.9212	0.9755	0.9942
50	0.1033	0.1132	0.1296	0.1524	0.1813	0.2161	0.2564	0.3014	0.3504	0.4026	0.4568	0.5121	0.5672	0.6212	0.7215	0.8068	0.8739	0.9228	0.9762	0.9944
60	0.1033	0.1132	0.1297	0.1526	0.1817	0.2167	0.2570	0.3022	0.3514	0.4038	0.4582	0.5136	0.5689	0.6229	0.7232	0.8084	0.8752	0.9238	0.9767	0.9946
80	0.1033	0.1133	0.1299	0.1529	0.1821	0.2173	0.2579	0.3033	0.3528	0.4053	0.4599	0.5155	0.5709	0.6250	0.7254	0.8103	0.8769	0.9251	0.9773	0.9948
100	0.1033	0.1134	0.1300	0.1531	0.1824	0.2177	0.2584	0.3039	0.3535	0.4062	0.4610	0.5167	0.5721	0.6263	0.7266	0.8115	0.8779	0.9258	0.9776	0.9949
120	0.1034	0.1134	0.1300	0.1532	0.1826	0.2179	0.2588	0.3044	0.3541	0.4068	0.4617	0.5174	0.5730	0.6271	0.7275	0.8123	0.8785	0.9263	0.9779	0.9950
200	0.1034	0.1135	0.1302	0.1534	0.1830	0.2185	0.2594	0.3053	0.3551	0.4081	0.4631	0.5190	0.5746	0.6288	0.7292	0.8138	0.8798	0.9273	0.9783	0.9951

$$\alpha = 0.20$$

f_2 \ ϕ =	0.1	0.2	0.3	0.4	0.5	0.6	0.7	0.8	0.9	1.0	1.1	1.2	1.3	1.4	1.6	1.8	2.0	2.2	2.6	3.0
1	0.2019	0.2074	0.2165	0.2290	0.2446	0.2629	0.2835	0.3062	0.3305	0.3561	0.3825	0.4094	0.4367	0.4639	0.5177	0.5694	0.6183	0.6638	0.7442	0.8102
2	0.2029	0.2114	0.2255	0.2448	0.2689	0.2972	0.3294	0.3646	0.4023	0.4419	0.4825	0.5236	0.5646	0.6050	0.6817	0.7508	0.8105	0.8599	0.9298	0.9687
3	0.2033	0.2133	0.2297	0.2521	0.2801	0.3130	0.3502	0.3907	0.4340	0.4789	0.5248	0.5706	0.6158	0.6596	0.7405	0.8101	0.8666	0.9101	0.9641	0.9879
4	0.2036	0.2144	0.2320	0.2562	0.2863	0.3216	0.3614	0.4047	0.4507	0.4983	0.5466	0.5945	0.6414	0.6863	0.7681	0.8363	0.8899	0.9295	0.9751	0.9928
5	0.2038	0.2150	0.2335	0.2587	0.2901	0.3269	0.3683	0.4133	0.4609	0.5101	0.5597	0.6088	0.6564	0.7019	0.7836	0.8507	0.9022	0.9393	0.9801	0.9948
6	0.2039	0.2155	0.2345	0.2604	0.2927	0.3305	0.3730	0.4191	0.4678	0.5179	0.5683	0.6181	0.6662	0.7119	0.7935	0.8597	0.9097	0.9450	0.9829	0.9958
7	0.2040	0.2158	0.2352	0.2617	0.2946	0.3332	0.3764	0.4233	0.4727	0.5235	0.5745	0.6247	0.6732	0.7190	0.8004	0.8658	0.9147	0.9488	0.9846	0.9964
8	0.2040	0.2161	0.2357	0.2626	0.2960	0.3351	0.3789	0.4264	0.4764	0.5276	0.5791	0.6296	0.6783	0.7242	0.8054	0.8702	0.9182	0.9515	0.9858	0.9968
9	0.2041	0.2162	0.2362	0.2634	0.2971	0.3367	0.3809	0.4288	0.4792	0.5309	0.5826	0.6334	0.6822	0.7282	0.8092	0.8735	0.9209	0.9535	0.9866	0.9970
10	0.2041	0.2164	0.2365	0.2640	0.2980	0.3379	0.3825	0.4308	0.4815	0.5335	0.5855	0.6364	0.6853	0.7313	0.8122	0.8761	0.9230	0.9550	0.9873	0.9972
12	0.2042	0.2166	0.2370	0.2649	0.2994	0.3398	0.3849	0.4337	0.4849	0.5373	0.5897	0.6409	0.6900	0.7360	0.8166	0.8799	0.9260	0.9571	0.9882	0.9975
14	0.2042	0.2168	0.2374	0.2655	0.3004	0.3411	0.3866	0.4358	0.4874	0.5401	0.5927	0.6441	0.6932	0.7393	0.8196	0.8825	0.9280	0.9586	0.9888	0.9977
16	0.2043	0.2169	0.2377	0.2660	0.3011	0.3421	0.3879	0.4373	0.4892	0.5421	0.5949	0.6464	0.6956	0.7417	0.8219	0.8844	0.9295	0.9597	0.9892	0.9978
18	0.2043	0.2170	0.2379	0.2664	0.3017	0.3429	0.3889	0.4386	0.4906	0.5437	0.5967	0.6483	0.6975	0.7436	0.8237	0.8859	0.9307	0.9605	0.9895	0.9979
20	0.2043	0.2171	0.2381	0.2667	0.3021	0.3435	0.3897	0.4395	0.4917	0.5450	0.5980	0.6497	0.6990	0.7451	0.8250	0.8871	0.9316	0.9611	0.9898	0.9980
22	0.2043	0.2172	0.2382	0.2669	0.3025	0.3440	0.3903	0.4403	0.4926	0.5460	0.5992	0.6509	0.7002	0.7463	0.8262	0.8880	0.9323	0.9617	0.9900	0.9980
24	0.2043	0.2172	0.2383	0.2671	0.3028	0.3444	0.3909	0.4410	0.4934	0.5469	0.6001	0.6519	0.7012	0.7473	0.8271	0.8888	0.9329	0.9621	0.9901	0.9980
26	0.2043	0.2173	0.2384	0.2673	0.3030	0.3448	0.3913	0.4415	0.4941	0.5476	0.6009	0.6527	0.7021	0.7481	0.8279	0.8894	0.9334	0.9624	0.9902	0.9981
28	0.2044	0.2173	0.2385	0.2674	0.3033	0.3451	0.3917	0.4420	0.4946	0.5482	0.6016	0.6534	0.7028	0.7489	0.8285	0.8900	0.9338	0.9627	0.9904	0.9981
30	0.2044	0.2173	0.2386	0.2676	0.3035	0.3453	0.3921	0.4424	0.4951	0.5488	0.6022	0.6541	0.7035	0.7495	0.8291	0.8905	0.9342	0.9630	0.9905	0.9981
35	0.2044	0.2174	0.2388	0.2678	0.3039	0.3459	0.3928	0.4433	0.4961	0.5499	0.6033	0.6553	0.7047	0.7507	0.8303	0.8914	0.9349	0.9635	0.9906	0.9982
40	0.2044	0.2175	0.2389	0.2680	0.3042	0.3463	0.3933	0.4439	0.4968	0.5507	0.6042	0.6562	0.7056	0.7517	0.8311	0.8921	0.9354	0.9639	0.9908	0.9982
50	0.2044	0.2175	0.2390	0.2683	0.3046	0.3468	0.3940	0.4448	0.4978	0.5518	0.6054	0.6575	0.7070	0.7530	0.8323	0.8931	0.9362	0.9644	0.9910	0.9983
60	0.2044	0.2176	0.2391	0.2685	0.3048	0.3472	0.3945	0.4453	0.4985	0.5525	0.6062	0.6583	0.7078	0.7538	0.8331	0.8938	0.9367	0.9647	0.9911	0.9983
80	0.2044	0.2177	0.2393	0.2687	0.3052	0.3477	0.3951	0.4461	0.4993	0.5535	0.6072	0.6594	0.7089	0.7549	0.8341	0.8946	0.9373	0.9651	0.9913	0.9984
100	0.2044	0.2177	0.2393	0.2688	0.3054	0.3480	0.3954	0.4465	0.4998	0.5540	0.6078	0.6600	0.7095	0.7555	0.8346	0.8951	0.9377	0.9654	0.9913	0.9984
120	0.2045	0.2177	0.2394	0.2689	0.3055	0.3481	0.3957	0.4468	0.5002	0.5544	0.6083	0.6605	0.7100	0.7560	0.8350	0.8954	0.9379	0.9656	0.9914	0.9984
200	0.2045	0.2178	0.2395	0.2691	0.3058	0.3485	0.3961	0.4474	0.5008	0.5552	0.6091	0.6613	0.7108	0.7568	0.8358	0.8960	0.9384	0.9659	0.9915	0.9984

3

Additional Noncentral F-Tables

Selected tables for the noncentral F-distribution with 1 and f_2 degrees of freedom. Values of ϕ as a function of $1 - \beta$ and f_2 for values of $\alpha = 0.02$, 0.05, 0.10, and 0.20. Values of ϕ are associated with two-tailed tests at α or one-tailed tests at $\alpha/2$.

α = 0.02

f_2 \ $1-\beta$	0.99	0.98	0.95	0.90	0.85	0.80	0.75	0.70	0.65	0.60	0.55	0.50	0.40	0.30	0.20	0.10
1	>7.00	>7.00	>7.00	>7.00	>7.00	>7.00	>7.00	>7.00	>7.00	>7.00	>7.00	>7.00	>7.00	>7.00	5.7032	2.8288
2	>7.00	>7.00	>7.00	>7.00	6.8845	6.3350	5.8734	5.4675	5.0991	4.7569	4.4333	4.1223	3.5199	2.9149	2.2638	1.4664
3	6.6026	6.1369	5.4499	4.8529	4.4581	4.1495	3.8886	3.6576	3.4464	3.2487	3.0601	2.8770	2.5161	2.1427	1.7230	1.1711
4	5.3176	4.9653	4.4438	3.9882	3.6853	3.4474	3.2454	3.0657	2.9006	2.7453	2.5964	2.4511	2.1621	1.8588	1.5118	1.0435
5	4.7257	4.4241	3.9763	3.5835	3.3214	3.1148	2.9388	2.7818	2.6372	2.5008	2.3695	2.2411	1.9845	1.7133	1.4004	0.9730
6	4.3930	4.1191	3.7115	3.3529	3.1130	2.9234	2.7616	2.6170	2.4836	2.3576	2.2361	2.1170	1.8784	1.6253	1.3318	0.9287
7	4.1822	3.9254	3.5427	3.2051	2.9789	2.7998	2.6469	2.5100	2.3835	2.2639	2.1485	2.0353	1.8081	1.5664	1.2855	0.8982
8	4.0376	3.7922	3.4261	3.1027	2.8856	2.7137	2.5667	2.4350	2.3133	2.1980	2.0868	1.9776	1.7581	1.5244	1.2522	0.8760
9	3.9326	3.6954	3.3410	3.0278	2.8172	2.6504	2.5076	2.3796	2.2613	2.1492	2.0410	1.9346	1.7208	1.4929	1.2271	0.8592
10	3.8530	3.6219	3.2763	2.9706	2.7649	2.6019	2.4622	2.3371	2.2213	2.1116	2.0056	1.9014	1.6919	1.4683	1.2075	0.8460
12	3.7408	3.5180	3.1846	2.8892	2.6903	2.5325	2.3974	2.2761	2.1639	2.0575	1.9547	1.8536	1.6501	1.4328	1.1789	0.8266
14	3.6656	3.4482	3.1227	2.8341	2.6397	2.4854	2.3532	2.2345	2.1247	2.0205	1.9198	1.8207	1.6213	1.4082	1.1591	0.8131
16	3.6117	3.3981	3.0782	2.7944	2.6032	2.4513	2.3212	2.2044	2.0962	1.9936	1.8944	1.7968	1.6002	1.3902	1.1445	0.8031
18	3.5713	3.3605	3.0447	2.7645	2.5758	2.4256	2.2969	2.1815	2.0746	1.9732	1.8751	1.7786	1.5842	1.3764	1.1333	0.7954
20	3.5399	3.3312	3.0186	2.7411	2.5540	2.4054	2.2780	2.1636	2.0576	1.9571	1.8599	1.7643	1.5716	1.3658	1.1245	0.7894
22	3.5147	3.3078	2.9976	2.7223	2.5366	2.3892	2.2627	2.1491	2.0440	1.9442	1.8477	1.7528	1.5614	1.3568	1.1174	0.7844
24	3.4942	3.2886	2.9805	2.7069	2.5224	2.3758	2.2501	2.1373	2.0327	1.9336	1.8376	1.7433	1.5530	1.3496	1.1115	0.7804
26	3.4771	3.2726	2.9662	2.6940	2.5105	2.3647	2.2396	2.1274	2.0233	1.9247	1.8292	1.7353	1.5460	1.3435	1.1066	0.7769
28	3.4626	3.2591	2.9540	2.6831	2.5004	2.3552	2.2307	2.1189	2.0153	1.9171	1.8220	1.7285	1.5400	1.3383	1.1023	0.7740
30	3.4502	3.2475	2.9436	2.6737	2.4917	2.3471	2.2231	2.1117	2.0085	1.9106	1.8159	1.7227	1.5348	1.3339	1.0987	0.7715
35	3.4258	3.2247	2.9231	2.6553	2.4746	2.3311	2.2079	2.0973	1.9949	1.8977	1.8037	1.7111	1.5246	1.3251	1.0915	0.7665
40	3.4078	3.2079	2.9080	2.6416	2.4620	2.3192	2.1967	2.0867	1.9848	1.8882	1.7946	1.7026	1.5170	1.3185	1.0861	0.7628
50	3.3832	3.1848	2.8872	2.6229	2.4445	2.3028	2.1813	2.0721	1.9709	1.8750	1.7821	1.6907	1.5065	1.3094	1.0787	0.7576
60	3.3670	3.1696	2.8736	2.6105	2.4331	2.2921	2.1711	2.0625	1.9618	1.8663	1.7739	1.6829	1.4996	1.3034	1.0738	0.7541
80	3.3472	3.1510	2.8568	2.5953	2.4190	2.2788	2.1585	2.0506	1.9505	1.8555	1.7637	1.6733	1.4910	1.2960	1.0677	0.7499
100	3.3355	3.1400	2.8468	2.5863	2.4106	2.2709	2.1511	2.0435	1.9438	1.8492	1.7576	1.6675	1.4859	1.2916	1.0640	0.7473
120	3.3278	3.1328	2.8403	2.5804	2.4051	2.2657	2.1462	2.0388	1.9393	1.8449	1.7536	1.6637	1.4825	1.2886	1.0616	0.7456
200	3.3125	3.1184	2.8273	2.5686	2.3941	2.2554	2.1364	2.0295	1.9305	1.8366	1.7458	1.6562	1.4758	1.2828	1.0568	0.7423

α = 0.05

f_2 \ $1-\beta$	0.99	0.98	0.95	0.90	0.85	0.80	0.75	0.70	0.65	0.60	0.55	0.50	0.40	0.30	0.20	0.10
1	>7.00	>7.00	>7.00	>7.00	>7.00	>7.00	>7.00	>7.00	>7.00	>7.00	6.8081	6.0788	4.7261	3.4727	2.2830	1.0951
2	6.8342	6.2926	5.4954	4.8052	4.3510	3.9976	3.7003	3.4384	3.2002	2.9786	2.7683	2.5658	2.1710	1.7698	1.3276	0.7447
3	4.8684	4.5172	3.9980	3.5453	3.2449	3.0094	2.8097	2.6323	2.4697	2.3169	2.1707	2.0282	1.7454	1.4494	1.1104	0.6387
4	4.2026	3.9136	3.4846	3.1083	2.8573	2.6595	2.4910	2.3407	2.2023	2.0718	1.9462	1.8233	1.5777	1.3178	1.0163	0.5894
5	3.8813	3.6212	3.2341	2.8933	2.6650	2.4847	2.3307	2.1931	2.0660	1.9459	1.8301	1.7166	1.4888	1.2467	0.9642	0.5612
6	3.6951	3.4513	3.0876	2.7666	2.5512	2.3808	2.2350	2.1045	1.9839	1.8697	1.7596	1.6514	1.4341	1.2024	0.9314	0.5430
7	3.5746	3.3410	2.9921	2.6836	2.4763	2.3121	2.1715	2.0456	1.9291	1.8188	1.7122	1.6075	1.3970	1.1722	0.9087	0.5304
8	3.4906	3.2639	2.9250	2.6251	2.4234	2.2634	2.1264	2.0036	1.8900	1.7823	1.6783	1.5761	1.3703	1.1504	0.8923	0.5212
9	3.4288	3.2071	2.8754	2.5817	2.3840	2.2271	2.0928	1.9723	1.8607	1.7550	1.6528	1.5524	1.3501	1.1338	0.8797	0.5141
10	3.3815	3.1636	2.8373	2.5482	2.3536	2.1991	2.0667	1.9479	1.8380	1.7338	1.6330	1.5339	1.3343	1.1208	0.8699	0.5085
12	3.3140	3.1012	2.7827	2.5001	2.3097	2.1586	2.0290	1.9127	1.8050	1.7029	1.6042	1.5070	1.3113	1.1018	0.8554	0.5002
14	3.2681	3.0589	2.7454	2.4671	2.2796	2.1307	2.0030	1.8884	1.7823	1.6816	1.5842	1.4884	1.2953	1.0885	0.8453	0.4944
16	3.2350	3.0282	2.7183	2.4432	2.2577	2.1104	1.9841	1.8707	1.7656	1.6660	1.5696	1.4747	1.2835	1.0787	0.8378	0.4901
18	3.2100	3.0050	2.6978	2.4250	2.2410	2.0949	1.9696	1.8571	1.7529	1.6540	1.5584	1.4643	1.2745	1.0712	0.8320	0.4868
20	3.1904	2.9869	2.6817	2.4107	2.2279	2.0828	1.9582	1.8464	1.7429	1.6446	1.5495	1.4560	1.2674	1.0653	0.8275	0.4842
22	3.1747	2.9722	2.6687	2.3992	2.2174	2.0729	1.9491	1.8378	1.7348	1.6370	1.5424	1.4493	1.2616	1.0605	0.8238	0.4820
24	3.1618	2.9603	2.6581	2.3897	2.2087	2.0648	1.9415	1.8307	1.7281	1.6307	1.5365	1.4438	1.2568	1.0565	0.8207	0.4803
26	3.1510	2.9502	2.6491	2.3817	2.2014	2.0581	1.9351	1.8247	1.7225	1.6254	1.5315	1.4391	1.2528	1.0532	0.8181	0.4788
28	3.1419	2.9417	2.6416	2.3750	2.1952	2.0523	1.9297	1.8197	1.7177	1.6209	1.5273	1.4352	1.2494	1.0503	0.8159	0.4775
30	3.1340	2.9344	2.6351	2.3692	2.1898	2.0473	1.9251	1.8153	1.7136	1.6171	1.5237	1.4318	1.2464	1.0478	0.8140	0.4764
35	3.1185	2.9200	2.6222	2.3577	2.1793	2.0375	1.9159	1.8066	1.7054	1.6094	1.5165	1.4250	1.2406	1.0429	0.8102	0.4742
40	3.1070	2.9093	2.6127	2.3492	2.1715	2.0302	1.9090	1.8002	1.6994	1.6037	1.5111	1.4200	1.2362	1.0393	0.8074	0.4725
50	3.0913	2.8946	2.5995	2.3374	2.1606	2.0201	1.8995	1.7913	1.6910	1.5958	1.5037	1.4130	1.2302	1.0342	0.8035	0.4703
60	3.0809	2.8849	2.5909	2.3297	2.1535	2.0134	1.8933	1.7854	1.6854	1.5905	1.4988	1.4084	1.2262	1.0309	0.8009	0.4688
80	3.0681	2.8729	2.5802	2.3201	2.1447	2.0052	1.8856	1.7781	1.6786	1.5841	1.4927	1.4027	1.2212	1.0267	0.7977	0.4669
100	3.0605	2.8659	2.5739	2.3144	2.1394	2.0003	1.8810	1.7738	1.6745	1.5802	1.4890	1.3993	1.2182	1.0242	0.7958	0.4658
120	3.0555	2.8612	2.5697	2.3107	2.1359	1.9970	1.8779	1.7709	1.6718	1.5777	1.4866	1.3970	1.2163	1.0226	0.7945	0.4650
200	3.0456	2.8519	2.5613	2.3032	2.1290	1.9906	1.8718	1.7652	1.6664	1.5726	1.4818	1.3925	1.2124	1.0193	0.7919	0.4635

$\alpha = 0.10$

f_2 \ $1-\beta$ =	0.99	0.98	0.95	0.90	0.85	0.80	0.75	0.70	0.65	0.60	0.55	0.50	0.40	0.30	0.20	0.10
1	>7.00	>7.00	>7.00	>7.00	6.5069	5.7928	5.1997	4.6848	4.2245	3.8043	3.4146	3.0488	2.3702	1.7382	1.1082	0.0000
2	4.8665	4.4761	3.9003	3.4006	3.0709	2.8136	2.5965	2.4046	2.2295	2.0659	1.9100	1.7589	1.4608	1.1501	0.7873	0.0000
3	3.8653	3.5776	3.1511	2.7777	2.5289	2.3330	2.1665	2.0180	1.8814	1.7526	1.6288	1.5076	1.2650	1.0064	0.6965	0.0000
4	3.5035	3.2513	2.8760	2.5455	2.3243	2.1494	2.0001	1.8666	1.7433	1.6267	1.5142	1.4038	1.1817	0.9431	0.6549	0.0000
5	3.3224	3.0873	2.7365	2.4266	2.2186	2.0538	1.9129	1.7867	1.6700	1.5594	1.4526	1.3476	1.1359	0.9078	0.6313	0.0000
6	3.2150	2.9896	2.6528	2.3548	2.1544	1.9955	1.8595	1.7375	1.6246	1.5177	1.4143	1.3125	1.1071	0.8854	0.6162	0.0000
7	3.1443	2.9251	2.5972	2.3068	2.1113	1.9563	1.8234	1.7043	1.5939	1.4893	1.3881	1.2885	1.0873	0.8700	0.6057	0.0000
8	3.0943	2.8794	2.5577	2.2726	2.0805	1.9281	1.7975	1.6803	1.5718	1.4688	1.3692	1.2711	1.0729	0.8587	0.5980	0.0000
9	3.0571	2.8453	2.5282	2.2469	2.0574	1.9070	1.7780	1.6622	1.5550	1.4533	1.3548	1.2579	1.0619	0.8500	0.5921	0.0000
10	3.0285	2.8190	2.5053	2.2270	2.0394	1.8905	1.7628	1.6481	1.5419	1.4411	1.3436	1.2475	1.0533	0.8432	0.5875	0.0000
12	2.9872	2.7811	2.4722	2.1981	2.0132	1.8664	1.7405	1.6275	1.5228	1.4234	1.3271	1.2323	1.0406	0.8332	0.5806	0.0000
14	2.9588	2.7550	2.4494	2.1781	1.9951	1.8498	1.7251	1.6132	1.5094	1.4110	1.3156	1.2217	1.0318	0.8262	0.5758	0.0000
16	2.9382	2.7360	2.4327	2.1635	1.9819	1.8376	1.7138	1.6026	1.4996	1.4018	1.3072	1.2139	1.0252	0.8210	0.5722	0.0000
18	2.9226	2.7215	2.4200	2.1523	1.9717	1.8282	1.7051	1.5946	1.4921	1.3948	1.3007	1.2079	1.0202	0.8170	0.5694	0.0000
20	2.9103	2.7101	2.4100	2.1435	1.9637	1.8208	1.6983	1.5882	1.4862	1.3893	1.2955	1.2031	1.0162	0.8138	0.5672	0.0000
22	2.9004	2.7010	2.4020	2.1364	1.9572	1.8149	1.6927	1.5830	1.4813	1.3848	1.2914	1.1993	1.0130	0.8113	0.5655	0.0000
24	2.8922	2.6934	2.3953	2.1305	1.9519	1.8099	1.6881	1.5788	1.4774	1.3811	1.2879	1.1961	1.0103	0.8091	0.5640	0.0000
26	2.8854	2.6871	2.3897	2.1256	1.9474	1.8058	1.6843	1.5752	1.4740	1.3780	1.2850	1.1934	1.0080	0.8073	0.5627	0.0000
28	2.8795	2.6817	2.3850	2.1214	1.9436	1.8023	1.6810	1.5721	1.4712	1.3753	1.2825	1.1911	1.0061	0.8058	0.5617	0.0000
30	2.8745	2.6771	2.3809	2.1178	1.9403	1.7992	1.6782	1.5695	1.4687	1.3730	1.2804	1.1891	1.0044	0.8045	0.5607	0.0000
35	2.8646	2.6679	2.3728	2.1106	1.9337	1.7932	1.6726	1.5642	1.4638	1.3685	1.2762	1.1852	1.0011	0.8018	0.5589	0.0000
40	2.8573	2.6611	2.3668	2.1053	1.9289	1.7887	1.6684	1.5603	1.4602	1.3651	1.2730	1.1823	0.9986	0.7999	0.5575	0.0000
50	2.8472	2.6517	2.3584	2.0979	1.9222	1.7824	1.6626	1.5549	1.4551	1.3603	1.2686	1.1782	0.9952	0.7971	0.5556	0.0000
60	2.8405	2.6455	2.3530	2.0931	1.9177	1.7783	1.6587	1.5513	1.4518	1.3572	1.2657	1.1755	0.9929	0.7953	0.5544	0.0000
80	2.8322	2.6378	2.3462	2.0870	1.9122	1.7732	1.6540	1.5469	1.4476	1.3533	1.2621	1.1721	0.9901	0.7930	0.5528	0.0000
100	2.8273	2.6332	2.3421	2.0834	1.9089	1.7702	1.6511	1.5442	1.4451	1.3510	1.2599	1.1701	0.9884	0.7917	0.5519	0.0000
120	2.8241	2.6302	2.3394	2.0810	1.9067	1.7682	1.6493	1.5425	1.4435	1.3495	1.2585	1.1688	0.9873	0.7908	0.5512	0.0000
200	2.8176	2.6242	2.3341	2.0763	1.9024	1.7641	1.6455	1.5390	1.4402	1.3464	1.2556	1.1661	0.9851	0.7890	0.5500	0.0000

225

$\alpha = 0.20$

f_2	$1-\beta=$ 0.99	0.98	0.95	0.90	0.85	0.80	0.75	0.70	0.65	0.60	0.55	0.50	0.40	0.30	0.20	0.10
1	5.8941	5.3233	4.4849	3.7638	3.2940	2.9325	2.6322	2.3714	2.1378	1.9238	1.7238	1.5335	1.1652	0.7733	0.0000	
2	3.4889	3.2011	2.7752	2.4034	2.1564	1.9623	1.7975	1.6506	1.5154	1.3876	1.2642	1.1426	0.8939	0.6090	0.0000	
3	3.0635	2.8227	2.4639	2.1479	1.9361	1.7686	1.6252	1.4968	1.3778	1.2647	1.1549	1.0461	0.8219	0.5621	0.0000	
4	2.8993	2.6754	2.3409	2.0451	1.8461	1.6883	1.5531	1.4316	1.3188	1.2115	1.1071	1.0035	0.7894	0.5406	0.0000	
5	2.8138	2.5983	2.2758	1.9901	1.7976	1.6448	1.5137	1.3958	1.2863	1.1820	1.0805	0.9797	0.7711	0.5283	0.0000	
6	2.7617	2.5511	2.2357	1.9560	1.7674	1.6176	1.4889	1.3732	1.2658	1.1633	1.0636	0.9645	0.7593	0.5204	0.0000	
7	2.7268	2.5193	2.2086	1.9328	1.7468	1.5990	1.4720	1.3578	1.2516	1.1504	1.0519	0.9540	0.7512	0.5149	0.0000	
8	2.7017	2.4965	2.1890	1.9160	1.7319	1.5854	1.4597	1.3465	1.2413	1.1410	1.0434	0.9463	0.7452	0.5108	0.0000	
9	2.6829	2.4793	2.1743	1.9033	1.7205	1.5752	1.4503	1.3379	1.2334	1.1339	1.0368	0.9404	0.7406	0.5077	0.0000	
10	2.6682	2.4660	2.1627	1.8934	1.7117	1.5671	1.4429	1.3312	1.2273	1.1282	1.0317	0.9358	0.7370	0.5053	0.0000	
12	2.6469	2.4464	2.1459	1.8789	1.6986	1.5553	1.4321	1.3212	1.2182	1.1199	1.0241	0.9289	0.7316	0.5016	0.0000	
14	2.6321	2.4329	2.1342	1.8687	1.6895	1.5470	1.4245	1.3143	1.2118	1.1140	1.0188	0.9241	0.7279	0.4991	0.0000	
16	2.6213	2.4230	2.1256	1.8613	1.6828	1.5409	1.4189	1.3091	1.2071	1.1097	1.0149	0.9205	0.7251	0.4972	0.0000	
18	2.6130	2.4154	2.1190	1.8555	1.6777	1.5362	1.4146	1.3052	1.2034	1.1064	1.0118	0.9178	0.7229	0.4957	0.0000	
20	2.6065	2.4094	2.1137	1.8510	1.6736	1.5325	1.4112	1.3021	1.2005	1.1037	1.0094	0.9156	0.7212	0.4945	0.0000	
22	2.6012	2.4045	2.1095	1.8473	1.6703	1.5295	1.4085	1.2995	1.1982	1.1016	1.0075	0.9138	0.7198	0.4936	0.0000	
24	2.5968	2.4005	2.1060	1.8443	1.6676	1.5270	1.4062	1.2974	1.1963	1.0998	1.0058	0.9124	0.7187	0.4928	0.0000	
26	2.5932	2.3971	2.1031	1.8417	1.6653	1.5249	1.4042	1.2958	1.1946	1.0983	1.0045	0.9111	0.7177	0.4921	0.0000	
28	2.5901	2.3943	2.1006	1.8396	1.6633	1.5231	1.4026	1.2941	1.1932	1.0970	1.0033	0.9101	0.7169	0.4916	0.0000	
30	2.5874	2.3918	2.0984	1.8377	1.6616	1.5215	1.4012	1.2928	1.1920	1.0959	1.0023	0.9092	0.7162	0.4911	0.0000	
35	2.5820	2.3869	2.0941	1.8339	1.6583	1.5185	1.3983	1.2902	1.1896	1.0937	1.0003	0.9073	0.7147	0.4901	0.0000	
40	2.5781	2.3832	2.0909	1.8312	1.6557	1.5162	1.3962	1.2883	1.1878	1.0921	0.9988	0.9060	0.7137	0.4894	0.0000	
50	2.5726	2.3782	2.0865	1.8273	1.6523	1.5130	1.3933	1.2855	1.1854	1.0898	0.9967	0.9041	0.7122	0.4883	0.0000	
60	2.5689	2.3748	2.0836	1.8247	1.6499	1.5109	1.3914	1.2838	1.1837	1.0883	0.9953	0.9028	0.7112	0.4877	0.0000	
80	2.5644	2.3706	2.0799	1.8215	1.6471	1.5082	1.3889	1.2815	1.1817	1.0864	0.9936	0.9013	0.7100	0.4868	0.0000	
100	2.5618	2.3682	2.0778	1.8197	1.6454	1.5067	1.3875	1.2802	1.1804	1.0853	0.9926	0.9003	0.7092	0.4863	0.0000	
120	2.5600	2.3665	2.0763	1.8184	1.6442	1.5056	1.3865	1.2793	1.1796	1.0845	0.9919	0.8997	0.7087	0.4860	0.0000	
200	2.5564	2.3633	2.0735	1.8159	1.6420	1.5036	1.3846	1.2776	1.1780	1.0830	0.9905	0.8985	0.7078	0.4853	0.0000	

References

Abramsky, Z. (1978). Small mammal community ecology: Change in species diversity in response to manipulated productivity. *Oecologia, 34,* 113–123.

Ahlgren, C. E. (1966). Small mammals and reforestation following prescribed burning. *J. Forestry, 64,* 614–618.

Allen, D. M., and F. B. Cady (1982). *Analyzing Experimental Data by Regression,* Lifetime Learning Pub., Belmont, CA.

Anderson, T. J., and G. W. Barrett (1982). Effects of dried sewage sludge on meadow vole (*Microtus pennsulvanicus*) populations in two grassland communities. *J. Appl. Ecol., 19,* 759–772.

Andrzejewski, R. (1975). Supplementary food and the winter dynamics of bank vole populations. *Acta Theriologica, 20,* 23–40.

Barrett, G. W., and R. M. Darnell (1967). Effects of dimethoate on small mammal populations. *Am. Midl. Nat., 77,* 164–175.

Bartlett, M. S. (1947). The use of transformations. *Biometrics, 3,* 39–52.

Beal, G. (1939). Methods of estimating the population of insects in a field. *Biometrika, 30,* 422–439.

Beck, A. M., and R. J. Vogl (1972). The effects of spring burning on rodent populations in a brush prairie savanna. *J. Mammal., 53,* 336–346.

Behrens, W. V. (1929). Ein Beitrag für Fehlen-Berechnung bei wenigen Beobachtungen. *Landwirtschaftliche Jahrbüchen, 68,* 507–837.

Bliss, C. I. (1967). *Statistics in Bioloy,* Vol. I, McGraw-Hill, New York.

Boonstra, R., and C. J. Krebs (1977). A fencing experiment on a high-density population of *Microtus townsendii. Can. J. Zool., 55,* 1166–1175.

Cameron, G. N. (1977). Experimental species removal: Demographic responses by *Sigmodon hispidus* and *Reithrodontomys fulvescens J. Mammal., 58,* 488–506.

Chapman, D. G. (1951). Some properties of the hypergeometric distribution with applications to zoological censuses. *Univ. Calif. Pub. Stat., 1,* 131–180.

Chapman, D. G, and W. S. Overton (1966). Estimating and testing differences between population levels by the Schnabel estimation method. *J. Wildl. Management.*, *30*, 173–180.

Chitty, D., and E. Phipps (1966). Seasonal changes in survival in mixed populations of two species of vole. *J. Am. Ecol.*, *35*, 313–331.

Churchman, C. W. (1948). *Theory of Experimental Inference*, Macmillan, New York.

Cochran, W. G. (1964). Approximate significance levels of the Behrens-Fisher test. *Biometrics, 20*, 191–195.

Cochran, W. G. (1977). *Sampling Techniques*, 3rd ed., Wiley, New York.

Cochran, W. G., and G. M. Cox (1957). *Experimental Designs*, 2nd ed., Wiley, New York.

Conover, W. J. (1971). *Practical Nonparametric Statistics*, Wiley, New York.

Cormack, R. M. (1964). Estimates of survival from the sighting of marked animals. *Biometrika, 51*, 429–438.

Cormack, R. M. (1968). The statistics of capture–recapture methods. *Oceanogr. Mar. Biol. Ann. Rev.*, *6*, 455–506.

Cormack, R. M. (1979). Models for capture–recapture. In *Sampling Biological Populations*, Statistical Ecology Series, Vol. 5, R. M. Cormack, G. P. Patil, and D. S. Robson, eds., International Co-op Publ., Burtonsville, MD, 217–255.

Cormack, R. M. (1985). Examples of the use of GLIM to analyse capture–recapture studies. In *Statistics in Ornithology*, B. J. T. Morgan and P. M. North, eds., Springer-Verlag, New York, pp. 243–273.

Cox, D. R. (1958). *Planning of Experiments*, Wiley, New York.

Crowner, A. W., and G. W. Barrett (1979). Effect of fire on the small mammal component of an experimental grassland community. *J. Mammal.*, *60*, 803–813.

Darroch, J. N. (1958). The multiple-recapture census. I: Estimation of a closed population. *Biometrika, 45*, 343–359.

Darroch, J. N., and D. Ratcliff (1980). A note on capture–recapture estimation. *Biometrics, 36*, 149–153.

Das, M. N., and N. C. Giri (1979). *Design and Analysis of Experiments*, Wiley Eastern Limited, New Delhi.

Desy, E. A., and C. F. Thompson (1983). Effects of supplemental food on a *Microtus pennsylvanicus* population in central Illinois. *J. Anim. Ecol.*, *52*, 27–70.

Eberhardt, L. L. (1976). Quantitative ecology and impact assessment, *J. Environ. Management, 4*, 27–70.

Eberhardt, L. L. (1978). Appraising variability in population studies. *J. Wildl. Management, 42*, 207–238.

Edgington, E. S. (1980). *Randomization Tests*, Marcel Dekker, New York.

Edwards, W. R., and L. L. Eberhardt (1967). Estimating cottontail abundance from live trapping data. *J. Wildl. Management, 31*, 87–96.

Federer, W. T. (1955). *Experimental Design: Theory and Application*, Oxford–IBH Publishing, New Delhi.

Federer, W. T. (1973). *Statistics and Society*, Marcel Dekker, New York.

Feller, W. (1971). *An Introduction to Probability Theory and Its Applications,* Vol. II, Wiley, New York.

Fisher, R. A. (1947). *The Design of Experiments,* 4th ed., Hafner, New York.

Fisher, R. A., and F. Yates (1957). *Statistical Tables for Biological, Agricultural and Medical Research,* 5th ed., Olives and Boyd, Edinburgh.

Fleiss, J. L. (1981). *Statistical Methods for Rates and Proportions,* Wiley, New York.

Flowerdew, J. R. (1972). The effect of supplementary food on a population of wood mice *(Apodemus sulvaticus). J. Anim. Ecol., 41,* 553–566.

Fordham, R. A. (1971). Field populations of deermice with supplemental food. *Ecology, 52,* 138–146.

Fox, B. J. (1982). Fire and mammalian secondary succession in an Australian coastal heath. *Ecology, 63,* 1332–1341.

Gaines, M. S., A. M. Vivas, and C. L. Barker (1979). An experimental analysis of dispersal in fluctuating vole populations: Demographic parameters. *Ecology, 60,* 814–828.

Gentry, J. G., R. B. Golley, and J. T. McGinnis (1966). Effect of weather on captures of small mammals. *Am. Midl. Nat., 75,* 526–530.

Getz, L. L. (1961). Response of small mammals to live traps and weather conditions. *Am. Midl. Nat., 66,* 160–170.

Gliwicz, J. (1970). Relation between trappability and age of individuals in a population of the bank vole. *Acta Theriologica, 15,* 15–23.

Grant, W. E., E. C. Birney, N. R. French, and D. M. Swift (1982). Structure and productivity of grassland small mammal communities related to grazing-induced changes in vegetative cover. *J. Mammal., 63,* 248–260.

Grant, W. E., N. R. French, and D. M. Swift (1977). Response of a small mammal community to water and nitrogen treatments in a shortgrass prairie ecosystem. *J. Mammal., 58,* 637–652.

Green, R. H. (1979). *Sampling Design and Statistical Methods for Environmental Biologists,* Wiley, New York.

Halvorson, C. H. (1982). Rodent occurrence, habitat disturbance, and seed fall in a larch-fir forest. *Ecology, 63,* 423–433.

Hansen, L., and G. O. Batzli (1978). The influence of food availability on the white-footed mouse: Populations in isolated woodlots. *Can. J. Zool., 56,* 2530–2541.

Hansen, M. H., W. N. Hurwitz, and W. G. Madow (1953). *Sampling Survey Methods and Theory,* Vol. I, *Methods and Applications,* Wiley, New York.

Hollander, M., and D. A. Wolfe (1973). *Nonparameteric Statistical Methods,* Wiley, New York.

Hoover, E. F. (1973). Response of the Oregon creeping vole to the clearcutting of a Douglas-fir forest. *N. West Science, 47,* 256–264.

Huber, J. J. 1962. Trap response of confined cottontail populations. *J. Wildl. Management, 26,* 177–185.

Hurlbert, S. H. (1984). Pseudoreplication and the design of ecological field experiments. *Ecol. Monogr., 54,* 187–211.

Jessen, R. (1978). *Statistical Survey Techniques,* Wiley, New York.

Jolly, G. M. (1965). Explicit estimates from capture–recapture data with both death and immigration-stochastic model. *Biometrika, 52,* 225–247.

Joule, J., and D. L. Jameson (1972). Experimental manipulation of population density in three sympatric rodents. *Ecology, 53,* 653–660.

Kirk, R. E. (1982). *Experimental Design: Procedures for the Behavioral Sciences,* Brooks/Cole, Belmont, CA.

Klonglan, E. D. (1955). Factors influencing the fall roadside pheasant census in Iowa. *J. Wildl. Management, 19,* 254–262.

Krebs, C. J., and K. T. DeLong (1965). A *Microtus* population with supplemental food. *J. Mammal., 46,* 566–573.

Krebs, C. J., B. L. Keller, and R. H. Tamarin (1969). *Microtus* population biology: Demographic changes in fluctuating populations of *M. ochrogaster* and *M. pennsylvanicus* in southern Indiana. *Ecology, 50,* 587–607.

Kremers, W. K. (1984). The Jolly–Seber model and maximum likelihood estimation revisited. BU-847-M. Biometrics Unit, Cornell University, Ithaca, NY.

Langley, A. K., Jr., and D. J. Shure (1980). The effects of loblolly pine plantations on small mammal populations. *Am. Midl. Nat., 103,* 59–65.

Lapin, L. (1975). *Statistics: Meaning and Method,* Harcourt Brace Jovanovich, New York.

Lee, F. S. (1972). Sample size determination for mark–recapture experiments. M.S. thesis, Cornell University, Ithaca, NY.

Lettenmaier, D. P., K. W. Hipel, and A. I. McLeod (1978). Assessment of environmental impacts, Part two: Data collection. *Environ. Management, 2,* 537–554.

LoBue, J., and R. M. Darnell (1959). Effects of habitat disturbance on a small mammal population. *J. Mammal., 40,* 425–437.

Manly, B. F. J., and M. J. Parr (1968). A new method of estimating population size, survivorship, and birth rate from capture–recapture data. *Trans. Soc. Br. Ent., 18,* 81–89.

May, R. M. (1981). *Theoretical Ecology: Principle and Applications,* Blackwell Scientific Pub., Oxford, England.

McKenzie, D. H., L. D. Kannberg, K. L. Gore, E. M. Arnold, and D.G. Watson (1977). Design and analysis of aquatic monitoring programs at nuclear power plants. PNL-2423. Battelle, Pacific Northwest Laboratories, Richland, WA.

Morrison, D. F. (1976). *Multivariate statistical methods,* McGraw-Hill, New York.

Munger, J. C., and J. H. Brown (1981). Competition in desert rodents: An experiment with semipermeable exclosures. *Science, 211,* 510–512.

Mystkowska, E. T., and J. Sidorowicz (1961). Influence of the weather on capture of micro-mammalia. II. Insectivora. *Acta Theriol., 5,* 263–273.

Nichols, J. D., and K. H. Pollock (1983). Estimation methodology in contemporary small mammal capture–recapture studies. *J. Mammal., 64,* 253–260.

O'Meara, T. E., J. B. Haufler, L. H. Stelter, and J. G. Nacy (1981). Nongame wildlife responses to chaining of pinyon-juniper woodlands. *J. Wildl. Management, 45,* 381–389.

Otis, D. L., K. P. Burnham, G. C. White, and D. R. Anderson (1978). Statistical inference for capture data from closed populations. *Wildlife Monograph 62.*

Paulik, G. J., and D. S. Robson (1969). Statistical calculations for change-in-ratio estimators of population parameters. *J. Wildl. Management, 33,* 1–27.

Perry, H. R., Jr., G. B. Pardue, F. S. Barkalow, Jr., and R. J. Monroe (1977). Factors affecting trap response of the gray squirrel. *J. Wildl. Management, 41,* 135–143.

Pollock, K. H. (1982). A capture–recapture design robust to unequal probability of capture. *J. Wildl. Management, 46,* 752–756.

Pollock, K. H., J. D. Nichols, C. Brownie, and J. E. Hines (1990). Statistical inference for capture–recapture experiments. *Wildlife Monograph 107.*

Pomeroy, S. E., and G. W. Barrett (1975). Dynamics of enclosed small mammal populations in relation to an experimental pesticide application. *Am. Midl. Nat., 93,* 91–106.

Press, W. H., B. P. Flannery, S. A. Teukolsky, and W. T. Vetterling (1986). *Numerical Recipes, The Art of Scientific Computing,* Cambridge Univ. Press, Cambridge, England.

Ricker, W. E. (1958). Handbook of computations for biological statistics of fish populations. *Bull. Fish. Board Canada, 119,* 1–300.

Robson, D. S. (1969). Mark–recapture methods of population estimation. In *New Developments in Survey Sampling,* N. L. Johnson and H. Smith, Jr., eds., Wiley-Interscience, New York, pp. 120–140.

Robson, D. S. (1974). On the possibility of exploiting autocorrelations within rows to estimate genetic and environmental variance and covariance in a plantation. *Commun. Statist., 3,* 913–922.

Robson, D. S., and H. A. Regier (1964). Sampling size in Petersen mark–recapture experiments. *Trans. Am. Fish. Soc., 93,* 215–226.

Romesburg, H. C. (1981). Wildlife science: Gaining reliable knowledge. *J. Wildl. Management, 45,* 293–313.

Rosenzweig, M. L. (1973). Habitat selection experiments with a pair of coexisting heteromyid rodent species. *Ecology, 54,* 111–117.

Scheffé, H. (1959). *The Analysis of Variance,* Wiley, New York.

Schnabel, Z. E. (1938). The estimation of the total fish population of a lake. *Am. Math. Monthly, 45,* 348–352.

Schroder, G. D., and M. L. Rosenzweig (1975). Perturbation analysis of competition and overlap in habitat utilization between *Dipodomus ordii* and *Dipodomys merriami. Oecologia, 19,* 9–28.

Searle, S. R. (1971). *Linear Models,* Wiley, New York.

Seber, G. A. F. (1965). A note on the multiple recapture census. *Biometrika, 52,* 249–259.

Seber, G. A. F. (1982). *The Estimation of Animal Abundance and Related Parameters,* Macmillan, New York.

Skalski, J. R. (1984). Towards the use of faunal populations as cost-effective biomonitors. In *Issues and Technology in the Management of Impacted Western Wildlife,* Thorne Ecological Institute, Boulder, CO, pp. 76–83.

Skalski, J. R. (1985a). Use of capture data to quantify change and test for effects on the abundance of wild populations. Ph.D. dissertation, Cornell University, Ithaca, NY.

Skalski, J. R. (1985b). Construction of cost functions for tag-recapture research. *Wild. Soc. Bull., 13*, 273–283.

Skalski, J. R., and D. H. McKenzie (1982). A design for aquatic monitoring programs. *J. Environ. Management, 14*, 237–251.

Skalski, J. R., and D. S. Robson (1979). Tests of homogeneity and goodness-of-fit to a truncated geometric model for removal sampling. In *Sampling Biological Populations*, R. M. Cormack, G. P. Patil, and D. S. Robson, eds., International Co-op. Pub. House, Burtonsville, MD, pp. 283–313.

Skalski, J. R., and D. S. Robson (1982). A mark and removal field procedure for estimating population abundance. *J. Wildl. Management, 46*, 741–751.

Skalski, J. R., D. S. Robson, and M. A. Simmons (1983a). Comparative census procedures using single mark–recapture methods. *Ecology, 65*, 1006–1015.

Skalski, J. R., D. S. Robson, and C. L. Matsuzaki (1983b). Competing probabilistic models for catch–effort relationship in wildlife censuses. *Ecol. Model., 19*, 299–307.

Skalski, J. R., M. A. Simmons, and D. S. Robson (1984). The use of removal sampling in comparative censuses. *Ecology, 65*, 1006–1015.

Smith, H. F. (1938). An empirical law describing heterogeneity in yields of agricultural crops. *J. Agric. Sci., 28*, 1–23.

Smith, M. H. (1968). A comparison of different methods of capturing and estimating numbers of mice. *J. Mammal., 49*, 455–462.

Smith, S. G., and J. R. Skalski (1989). Statistical design and analysis of avian field trials in environmental toxicology. *Pesticides in Terrestrial and Aquatic Environments*, D. L. Weigmann, ed., Virginia Water Resource Research Center, Blacksburg, VA, pp. 407–421.

Snedecor, G. W. (1956). *Statistical Methods Applied to Experiments in Agriculture and Biology*, Iowa St. College Press, Ames, IA.

Snedecor, G. W., and W. G. Cochran (1980). *Statistical Methods*, 7th ed., Iowa St. Univ. Press, Ames, IA.

Sprott, D. A. (1981). Maximum likelihood applied to a capture-recapture model. *Biometrics, 37*, 371–375.

Taitt, M. J., and C. J. Krebs (1981). The effect of extra food on small rodent populations. II. Voles (*Microtus townsendii*). *J. Anim. Ecol., 50*, 125–137.

Tamarin, R. H., and C. J. Krebs (1972). Selection at the transferrin locus in cropped vole populations. *Heredity, 30*, 53–62.

Taylor, L. R. (1961). Aggregation, variance and the mean. *Nature, 189*, 732–735.

Tester, J. R. (1965). Effects of a controlled burn on small mammals in a Minnesota oak-savanna. *Amer. Midl. Nat., 74*, 240–243.

Thomas, J. M., J. F. Cline, C. E. Cushing, M. C. McShane, J. E. Rogers, L. E. Rogers, J. C. Simpson, and J. R. Skalski (1983). Field evaluation of hazardous waste site bioassessment protocols. PNL-4614/UC-11. Pacific Northwest Laboratory, Richland, WA.

Welch, B. L. (1938). The significance of the difference between two means when the population variances are unequal. *Biometrika, 29*, 350–362.

White, G. C., D. R. Anderson, K. P. Burnham, and D. L. Otis (1982). Capture–recapture and removal methods for sampling closed populations. LA-8787-NERP, Los Alamos Natl. Lab, Los Alamos, NM.

Williams, D. A. (1971). A test for differences between treatment means when several dose levels are compared with a zero dose control. *Biometrics, 27,* 103–118.

Wilson, E. B., Jr. (1952). *An Introduction to Scientific Research,* McGraw-Hill, New York.

Wittes, J. T. (1974). Applications of a multinomial capture–recapture model to epidemiological data. *J. Am. Stat. Assoc., 69,* 93–97.

Young, H., J. Neess, and J. T. Emlen, Jr. (1952). Heterogeneity of trap response in a population of house mice. *J. Wildl. Management, 16,* 169–180.

Zippin, C. (1956). An evaluation of the removal method of estimating animal populations. *Biometrics, 12,* 163–169.

Zippin, C. (1958). The removal method of population estimation. *J. Wildl. Management, 22,* 82–90.

Index